GLOBAL THINKING AND LOC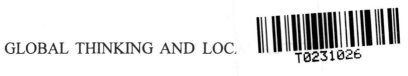

This book is dedicated to:

*My special friend, Margaret Moffat (Meg), for all her love
and affection;*

*My parents (Rev. Efiong A. Ite [1934-1997] and Deaconess Afiong E. Ite),
for their great devotion and sacrifice for our education;*

*My brothers (Victor, Emem, Obot, Aniefiok and Kufre), for their
listening ears and helping hands;*

and

*My sisters (Ini, Grace and Nsikak), for being so dependable
and caring.*

Global Thinking and Local Action

Agriculture, tropical forest loss and conservation in Southeast Nigeria

UWEM E. ITE
Department of Geography, Lancaster University, United Kingdom

Routledge
Taylor & Francis Group

LONDON AND NEW YORK

First published 2001 by Ashgate Publishing

Reissued 2018 by Routledge
2 Park Square, Milton Park, Abingdon, Oxon OX14 4RN
711 Third Avenue, New York, NY 10017, USA

Routledge is an imprint of the Taylor & Francis Group, an informa business

Publisher's Note
The publisher has gone to great lengths to ensure the quality of this reprint but points out that some imperfections in the original copies may be apparent.

Disclaimer
The publisher has made every effort to trace copyright holders and welcomes correspondence from those they have been unable to contact.

A Library of Congress record exists under LC control number: 00111421

ISBN 13: 978-1-138-70199-1 (hbk)
ISBN 13: 978-1-138-70202-8 (pbk)
ISBN 13: 978-1-315-20377-5 (ebk)

Contents

PART I: GLOBAL THINKING: THEORETICAL ISSUES

1 Introduction

2 Small Farmers, Tropical Forest Loss and Conservation

PART II: ENVIRONMENTAL MANAGEMENT ISSUES IN NIGERIA

3 Tropical Forest Loss and Resource Conservation in Nigeria

List of Figures

List of Tables

List of Abbreviations

CIFOR	Center for International Forestry Research
CRADP	Cross River Agricultural Development Project
CRFD	Cross River Forestry Department
CRNP	Cross River National Park
CRNPP	Cross River National Park Project
CRS	Cross River State
CRSG	Cross River State Government
CTZ	Conservation and Tourism Zone
CWD	Conservation-with-Development
EC	European Community
ECU	European Currency Unit
EDG	Environment and Development Group
ER	Eastern Region
ETFRN	European Tropical Forest Research Network
EU	European Union
FAO	Food and Agricultural Organisation
FDD	Forest Development Department
FEPA	Federal Environmental Protection Agency
FGN	Federal Government of Nigeria
FNPS	Federal National Parks Service
GCP	Gorilla Conservation Project
GDP	Gross Domestic Product
GERN	Government Eastern Region of Nigeria
HRH	His Royal Highness
ICDP	Integrated Conservation Development Project
IOH	Ikom-Obudu Highway
IPP	Intermediate Phase Project
IUCN	International Union for the Conservation of Nature and Natural Resources (The World Conservation Union)
KfW	*Kreditanstalt fur Weideraufban*
LGA	Local Government Area
MMCP	Mbe Mountains Conservation Project
MMSZDP	Mbe Mountains Support Zone Development Programme
MTR	Mid-term Review

NARESCON	Natural Resources Conservation Council
NCES	National Conservation Education Strategy
NCF	Nigerian Conservation Foundation
NCS	National Conservation Strategy
NEST	Nigeria Environmental Study/Action Team
NGO	Non-governmental Organisation
NTFP	Non-timber Forest Product
ODA	Overseas Development Administration
PA	Protected Area
SZ	Support Zone
SZDA	Support Zone Development Association
SZDP	Support Zone Development Programme
TMF	Tropical Moist Forest
T&V	Training and Visit
UNEP	United Nations Environment Programme
UK	United Kingdom
US	United States
VIRDC	Village Integrated Rural Development Committee
VLA	Village Liaison Assistant
WCED	World Commission on Environment and Development
WCMC	World Conservation Monitoring Centre
WCS	World Conservation Strategy
WID	Women in Development
WWF	World Wide Fund for Nature

Preface

This book focuses on the role of small farmers in tropical moist forest (TMF) loss, and the impact of forest conservation on farmers in southeast Nigeria. The central argument is that an understanding of smallholder forest farmers is vital to attempts to conserve TMF environments, particularly where 'conservation-with-development' or 'integrated conservation and development project' approaches are used. Without such knowledge, tropical forest conservation initiatives are unlikely to be successful.

There is no doubt that loss of TMF environment has been recognised as a major environmental problem globally and particularly in the West African sub-region. Although the significant causes of tropical forest loss are relatively clearly understood, the specific contribution of some causes, particularly the role of smallholder forest farmers, is less well known. Theoretically, the causes of TMF loss are commonly explained in the contexts of human ecology and political economy. Analyses based on these frameworks are often inadequate. This is because these perspectives do not provide sufficient insights on the sources and nature of conflicts arising from the competing interests in the management and development of local forest resources.

The role of forest farmers is known to be significant in forest loss in West Africa. However, the relevant underlying local processes have not been adequately understood or fully documented. Knowledge of tropical forest loss in the sub-region is almost entirely derived from aggregate national data. There has been little or no detailed local (micro or village level) study and analysis of TMF loss. This is particularly true in Nigeria, where the problem of forest loss has reached an advanced stage. Yet such studies are necessary both for comprehending local patterns of forest loss and their links to broader patterns of human activity.

On the other hand, tropical forest conservation is now a high priority in West African countries and particularly in Nigeria. However, the conservation efforts in the sub-region are built on inadequate understanding of the specific processes of degradation or the way human populations and economy, society and natural systems in TMF environments interact. The result has been that conservation projects are being designed and implemented based on 'top-down' development strategies, which implies that the socio-economic impact of such projects on local resource-dependent communities can outweigh the potential benefits. There is therefore a need for a better understanding of the behaviour of small farmers with a view to articulating this knowledge in conservation policies.

This book contributes to the emerging debates about the role of smallholder farmers in TMF loss both specifically in West Africa and by implication elsewhere, and the way that forest conservation programmes can engage with forest farmers. It provides a detailed account of the tensions between small farmers, agriculture and tropical forest conservation in southeast Nigeria. Using the Okwanwgo Division of the Cross River National Park as the case study, the book explores the causes of TMF loss at the household level, and examines the conflicts over access due to competing interests. A 'bottom-up' approach is adopted, with the smallholder forest-farming household as the focal point of analysis. The book links local level agricultural practices and household decision-making with the wider political economy to explain the observed patterns of forest loss in the Division. By focusing on the dynamics of forest farming at the household level, the book makes the case for an alternative perspective on the role of small farmers in TMF loss in West Africa, to that of existing studies of the region.

Uwem E. Ite

Acknowledgements

This book has been developed out of my doctoral thesis at the Department of Geography, University of Cambridge, United Kingdom. Both the thesis and this book would never have been written without the support from my family, friends and colleagues who have made diverse and significant contributions to my academic achievements.

Speaking of academic achievements, my interest in the relationship between local people, tropical forests and National Parks in developing countries emerged in 1991 while I was undertaking a course of study leading to the award of the MPhil degree in Environment and Development at the University of Cambridge. Against this background, I am very grateful to Dr Bill Adams, whose inspiring lectures on sustainable development and his classic book (*Green Development*) opened my inquiring mind on the subject under review. His subsequent supervision of my MPhil dissertation (on interregional water transfer schemes) and PhD thesis (on tropical forest loss and conservation) subsequently provided the leverage for me to pursue an academic career within the framework of environment and development research.

I am also very grateful to Valerie Rose (Commissioning Editor, Ashgate Publishing) and Dr Chris Park (Lancaster University) who were the first to suggest the idea of transforming my PhD thesis into a book. However, it has to be acknowledged that the experiences and ideas expressed in this book have been significantly shaped through my contact with several tropical forestry and conservation professionals. They include Nick Ashton-Jones, John Barker, Richard Barnwell, Julian Caldecott, Robert Dunn, Donald Gordon, John Makong, Phil Marshall, Tunde Morakinyo and Caroline Okpei. As such, my sincere gratitude is also extended to them for sharing with me, their practical field experiences on some of the issues raised in this book.

The support and encouragement from many friends and acquaintances was significant and crucial to the completion of this book. I am particularly indebted to Dr Adekunle and Adefolake Adeyeye, for their excellent hospitality during my numerous visits to Cambridge. Similarly, I appreciate the moral support from Katherine Saunders, Rod and Alison Lamb, Jessica Faleiro and Mark Holmes, who frequently checked on my progress with preparing the book manuscript. Their collective gesture unknowingly made me put more effort into getting the draft manuscript ready within the shortest possible time.

I received very constructive comments on the book manuscript from colleagues and professional associates. As such, I am very thankful to Prof. Rick Auty, Prof. Graham Chapman, Dr Gordon Clark and Dr Chris Park; Dr Roger Blench (Overseas Development Institute, London), Tunde Morakinyo (Iroko Foundation, London) and Prof. John Oates (State University of New York, USA). As might be expected, I did not always agree with all their views and suggestions, in as much as they were very useful.

I am highly indebted to the organisations and institutions that provided the relevant project documentation and technical support for the research and preparation of this book. These include WWF-UK, Godalming; Cross River National Park (Okwangwo Division); Federal Surveys, Lagos; Cross River Forestry Department, Calabar; University of Cambridge and Lancaster University.

My sincere gratitude goes to our Senior Cartographer, Chris Beacock, who did an excellent job of producing the maps and the camera-ready copy (CRC) of this book. In the same vein, I cannot fail to thank Anne Keirby (Ashgate Publishing) for being extremely patient, especially when I was unable meet the various deadlines for submitting the CRC.

Finally, I accept full responsibility for any factual errors this book may contain.

Uwem E. Ite
Lancaster, June 2000

PART I:
GLOBAL THINKING:
THEORETICAL ISSUES

1 Introduction

1.1 Context

The loss of tropical moist forests (TMF) is widely accepted as a global environmental problem. Despite conflicting accounts of the global extent, rates and accuracy of estimates of TMF loss, there is no doubt that the remaining TMF cover in many countries represents a tiny fraction of the natural vegetation cover. It is evident that forests continue to disappear for agricultural purposes throughout the humid tropics, with both immediate and potentially large long-term consequences for climate change and loss of biodiversity, both of which are of significant environmental interest to the international community (Tomich *et al.*, 1999; Laurance, 1999). Within the context of developing countries, including those in sub-Saharan Africa, several authoritative sources attribute deforestation to the actions of the rural poor, while others argue that this is tantamount to blaming the victim and directing attention away from the fundamental causes of deforestation. This is so even though there is little agreement on the definition of deforestation used to test hypotheses on the problem (Rock, 1996). Furthermore, conventional wisdom and thinking about the causes of deforestation is increasingly and vigorously challenged (Angelsen and Kaimowitz, 1999). Yet, many TMF conservation projects have been designed and implemented without an adequate understanding of the specific local processes and causes of deforestation. This has significant implications for the livelihood systems of communities adjacent to protected forest areas, especially National Parks in developing countries.

1.2 National Parks and Local Communities

Since the 1980s, several initiatives have been adopted in many developing countries to check the problem of TMF loss. Some are local in organisation (Jain, 1984; Weber, 1988), while others are significantly international in orientation (Rubinoff, 1983; Guppy, 1983; Poore and Sayer, 1987). Efforts by international conservation groups and organisations to preserve TMF across the globe concentrated on the establishment of protected areas (PAs). The major objective of the global protected area system has always been the main-

3

tenance (preservation) of the diversity of species and ecosystems (IUCN, 1994a). PAs constitute one approach to TMF conservation. Inspite of the detailed and comprehensive system of protected area classification, the National Park has been the most widely known and frequently used category in developing countries (Brechin *et al.*, 1991).

According to Hannah (1992, p.4), the management objective of the National Park is:

> To protect natural and scenic areas of national or international significance for spiritual, educational, recreational and tourism purposes. The area should perpetuate, in a natural state, representative samples of physiographic regions, biotic communities, genetic resources, and species, and to provide ecological stability and diversity.

There is sufficient evidence to demonstrate that the creation of National Parks has often resulted in conflicts with the interests of local people who may have been surviving on the resources of the designated protected area (see Hough, 1988; Brechin *et al.*, 1991). By definition, designation of a protected area implies some restriction on use of its resources (Hales, 1989). The international definition of 'National Park' includes a requirement for the highest competent authority of the country to prevent or eliminate exploitation or occupation of the park area (IUCN, 1985). Eliminating exploitation normally precludes local people's access to traditionally used resources. Eliminating occupation means relocation for communities in the park, with its attendant potential for disastrous side effects (Hough, 1988). Furthermore, the control of National Parks by central or national government effectively presumes that only the national government can manage the land better than any local authority (McNeely, 1989). As a result, local view has often been ignored in the process.

As West and Brechin (1991) observed, some areas designated as PAs may already have been inhabited and managed by local or indigenous people justifiably exploiting the resources on a regular basis for subsistence. More often, the traditional approaches to government management of such areas relied on guard patrols and penalties to exclude local people, a strategy characterised (Wells *et al.*, 1992) as the 'fines-and-fences' approach. The root of this strategy stems from the fact that Western scientists and international development agencies have a strong influence on conservation policy in developing countries, including sub-Saharan Africa (Leach and Mearns, 1996). The philosophy behind such policies makes a distinction between what is 'natural' and what is 'man-made'. For example, WWF International (1991) asserted that the surest form of protection for TMF is 'isolation' from industrial society. However, in many TMF areas, humans are an integral part of the forest ecosystem. Therefore a distinction between what is 'natural' and what is 'man-made' becomes rather difficult to make in theory and totally impossible to achieve in practice, without serious conflicts arising.

Within the context of sub-Saharan Africa, the 'fines-and-fences' approach (i.e. the American National Park model) to wildlife protection is now perceived by many conservationists to have failed (Songorwa, 1999). Conservation policies based on such ideas do not consider the needs, aspirations and values of the people whose livelihood would be most affected by these plans (Lusigi, 1984 and Anderson and Grove, 1987). As Fletcher (1990) suggested, this is in keeping with the western tradition of keeping people away from a 'natural landscape'. In reality the frequent and excessive use of this category of PAs (i.e. National Park), and the emulation of the United States park system in developing countries has contributed significantly to the complexity of the relationship between local people and national parks in developing countries.

Unlike the British model of National Parks which has always included people (Dower, 1992), the United States National Park model is based on the notion of preserving vast pristine 'wilderness' void of human presence, an example of which is the Yellowstone National Park designated in the 1870s. The idea that people should not live in protected areas or exploit their resources is virtually synonymous with the US National Park ideal. In otherwords and from a structural perspective, the concept of the National Park as developed in the US and adopted internationally, was one of a large land area from which all human activities (except those associated with management and tourism) were excluded. It is important to note here that the first expression of the concept by George Catlin in 1832 did involve native Americans, although this vision was never adopted (Zube and Busch, 1990). Nonetheless, this idea represents early recognition of the importance of the role of local people in the Park, and their place in PA management. As Hales (1989) has noted, much of the energy and philosophy of the National Park development at the international level have been derived from its North American heritage, especially the relationship between local people and the National Park.

In sub-Saharan Africa, as elsewhere, there are critical debates on the definition and purpose of conservation. The key issues have been focused on whether to consider conservation as meaning the preservation of wildlife or a balance of people and natural resources. From a historical perspective (see Anderson and Grove, 1987; Grove, 1987, 1990; Lindsay, 1987; MacKenzie, 1987), it is clear that by the end of the 19th century, the purpose of official conservation was the preservation of individual species and places. Colonial governments imposed the conservation movement on sub-Saharan Africa

(Adams, 1990), where there was, and still exists, a fascination for the unique number of wildlife species. In a forceful exposition of the mythical principles underlying conservation strategies in Africa, Adams and McShane (1992, p.xviii) submitted that:

> Conservation has long been operated on the comfortable belief that Africa is a paradise to be defended, even against the people who have lived there for thousands of years. The continuing reluctance to accept the link between vigorous indigenous culture and the survival of wildlife has led to conservation programmes doomed to eventual failure because they depend on building barriers of one sort or another between people and wildlife.

Traditional protected area (especially National Parks) management was therefore generally unsympathetic to the needs of, and constraints facing, local people. Numerous examples exist in sub-Saharan Africa where the establishment of National Parks effectively removed resources from local use without consideration for the replacement of lost resources or the basic needs of the local population (McShane, 1990). For example, the use of wild plants and animals and other ecological resources has been a traditional feature of most African economies irrespective of the dominant means of sustenance (McShane, 1990). Nonetheless, in some National Parks, residents and local populations who had traditionally used the park area for physical subsistence or spiritual needs were excluded because park practices and policies rarely, if ever, provided for such uses (Zube and Busch, 1990). Increasingly, studies have shown that ignoring the dependence of local people on park resources for their subsistence needs and emphasising law enforcement aggravated conflicts between the local people and park managers (Sharma, 1990; Ite, 1996). The Amboseli National Park in Kenya is a particularly good example (Wells *et al.*, 1992).

Cases abound where people who had been living inside National Parks were either forcibly evicted or allowed to remain in small enclaves inside the Park boundaries. For example, the Rendille were excluded from the Sibili National Park in Kenya and the Ik expelled from the Kidekpo National Park in Uganda (MacKinnon *et al.*, 1986). In Namibia, the Bushmen in the Gemsbok National Park became the major objects of conservation due to management policies which restricted the extent of their interaction with 'foreigners' (Gordon, 1985). Turton (1987) questioned the rationale for the preference given to wildlife conservation in the Omo National Park in Ethiopia, rather than the survival of the Mursi people whose economy very much depends on the ecosystem dynamics of the Omo River.

Since the 1980s, oppressive National Park and conservation policies in developing countries have been seriously questioned. Local communities and their people objected to the harsh enforcement tactics. Several scholars decried the absence of local participation and the disregard for traditional practices (Anderson and Grove, 1987; Cartwright, 1991; Hannah, 1992; Newmark *et al.*, 1993). It is now widely held that understanding local culture, its traditional resource management practices, and its inherent attitudes towards the natural world should be a prerequisite to any nature conservation initiative. Local knowledge systems, cultural values and traditional resource-use practices have long formed the basis for sustainable resource management in many societies (Brown, 1991; Schelhas, 1991, 1992; Matowanyinka, 1992; Schelhas and Shaw, 1992). The above views reinforce the assertion that the establishment of protected natural areas can have adverse effects on local economies and cultures, if adequate attention is not given to the interactions between people and their natural environment (Dasmann, 1984). These concerns have led to significant progress towards the evolution and adoption of new strategies for protected area management.

1.3   Strategies for Protected Area Management

The negative consequences of the imposition of National Parks on rural communities are diverse. These range from restriction of access to resources, the disruption of local cultures and economies by tourists, increased depredations on crops and livestock by wild animals, and the displacement of peoples from their traditional lands. The results are evident in social and cultural disruption, enforced poverty and even death. These adverse effects generate resentment and hostility against protected area management (Hough, 1988).

The literature on bringing people and parks together dates back from the classic essay by Western (1982) on a new management approach for the Amboseli National Park in Kenya. Since then, several scholars have advocated new concepts that promised to integrate people into protected area planning. Various new National Park design strategies and management approaches have been proposed (see Schonewald-Cox and Bayless, 1986; Leader-Williams *et al.*, 1990; Shafer, 1999). These include less protectionist management policies permitting limited sustainable resource extraction from National Park, increased local involvement in tourism, increased park-related revenues received by local people, and the provision of alternative sources of income for national park neighbours (Songorwa, 1999). Buffer zones have been established around National Parks (Sayer 1991a), with land uses that can both increase a park's ability to protect biological resources and provide subsistence as well as income-generating opportunities for local people. Other initiatives include development projects outside National Parks, such as agricultural intensification and cultivation of substitute resources. These are aimed at reducing local peoples' need for, and dependence on, the resources within the National Park boundaries (Schelhas, 1992).

The relationship between local populations and PAs has received international attention from many researchers, park managers and international development interests (Zube and Busch, 1990). There is evidence to suggest that the conservation community has acknowledged that communities next to protected area boundaries frequently bear substantial costs through the loss of access to park resources (Wells *et al.*, 1992). There is also a growing realisation that the local communities and their lifestyles need protection just as much as endangered wildlife species and ecosystems; and that cultural diversity be protected as well as biodiversity. It is becoming increasingly clear that for conservation to become sustainable in the long-term, the interests of local people and conservationists must converge. To this end, an assessment of human attitudes and potential for actual conflicts is critically important in the planning for long-term conservation strategies within the context of protected area management. Although the conservation of biodiversity has generally been considered a public good, there is also growing recognition that the justification for the creation and maintenance of PAs (including National Parks) must include analyses of rural livelihoods (McNeely, 1988; Dixon and Sherman, 1990; Kramer *et al.*, 1994). Recent studies suggest two main issues: the need to define and respect the rights of local people and the practical need to involve them in reserve management lest they wreck the reserve.

There has been significant interest in the study of conflicts arising from the traditional approach to the establishment and management of PAs. Hough (1988 p.134) made a compelling case for a shift in the ruling paradigm of PAs by pointing out that:

> The concept of national parks as inviolate havens of untouched Nature, controlled by an all-powerful central government agency, will have to give way to concepts of conservation through careful manipulation to achieve both conservation and local human development objectives.

A strong consensus has emerged that a National Park must involve local people in management decisions; that local people must benefit from National Parks; and that the support of local people is essential to the long-term existence of PAs. In analysing past failures of conservation in sub-Saharan Africa, Lewis *et al.* (1990) suggested that conservation implemented solely by the government for the general benefit of the whole population is likely to have limited success, especially in economically weak countries. As Lado

(1992) noted, the involvement of local people and consideration of their immediate livelihood strategies, perceptions and values by policy makers and planners is also vital for the success of protected area management. It is evident that protected area management efforts have now shifted in philosophy from a confrontational approach between park and local people to one that seeks to address the needs of local communities. This approach has been described as 'conservation-with-development'.

1.4   Conservation-with-Development

Conservation-with-development (CWD) is a relatively new concept. It is a component of the wider debate on 'sustainable development', a term widely used in the political rhetoric of development (Stocking and Perkin, 1992). In simple terms, CWD is the product of the thinking that conservation is essential for development, and development a prerequisite for successful conservation. This was the cornerstone of the World Conservation Strategy (WCS), which placed significant emphasis on the importance of linking PA management with the economic activities of local communities (IUCN, 1980). It became the central theme of the 1982 World Park Congress (McNeely and Miller, 1983) and was later reiterated by several international environment and development policy documents, including the Brandt Report (Brant, 1983) and Brundtland Report (WCED, 1987). As Stocking and Perkin (1992, pp.338 and 399) have put it:

> CWD is an approach conceptually derived from conservationists and environmentalists alarmed by the accelerating loss of species and frustrated by the silence of development agencies and the antagonism of many local peoples....
> ... The problem with CWD, as indeed with sustainability, is that it was originally developed with a biological and physical focus. Conservationists saw it as a way of gaining greater attention for their primary goal: preservation of species and/or natural resources. The economic and social context was drawn only later, leading to considerable confusion and a lack of guidelines.

Several international conservation organisations and development agencies including the World Conservation Union (IUCN), World Wide Fund for Nature (WWF) and the United Nations Environment Programme (UNEP), have embraced this new philosophy (see, for example, Larson *et al.*, 1998). The result has been the design and implementation of what Wells *et al.* (1992) described as 'integrated conservation-development projects' (ICDPs) or what Hannah (1992) called 'people and park projects'. In an attempt to explicitly portray the linkages between conservation and development, several research-ers (e.g. Gradwhol and Greenberg, 1988) presented glowing and very opti-mistic descriptions of successful projects. In sub-Saharan Africa, for exam-ple, Mascarenhas (1983) described the Ngorongoro Conservation Area in Tanzania as a unique balancing act between conservation and development. Yet, ICDPs have been generally considered to have broad, and to a large extent, very ill defined components. In many respects, ICDPs have been shown to be ineffective in addressing the links between conservation and develop-ment (Stocking and Perkin, 1992).

An international survey by Zube and Busch (1990) identified four cat-egories or models of park-people relationship reflecting this new approach and philosophy in PA management. These are:

• Local people's participation in park management and operations.
• Delivery of services (e.g. education, health care) by park authorities to the local communities.
• Allowance for the continuation of traditional land uses e.g. hunting and gathering, agriculture, religious practices, pastoralism.
• Involvement of local people in park related tourism.

However, a closer examination of specific ICDPs will reveal signifi-cant differences in the results of adopting any or a combination of the above models. For example, a survey and socio-economic analysis of park-people relations in Kosi Tappu Wildlife Reserve, Nepal, found that most local people had negative attitudes about the Reserve, despite the direct and indirect ben-efits they had received (Heinen, 1993). This was largely attributed to the com-munities' overestimation of crop damage by animals and an underestimation of benefits derived from the management of the Reserve. On the other hand, a survey of public attitudes and needs around the Kasungu National Park, Malawi (Mkanda and Munthali, 1994) showed that respondents' attitudes to-wards the conservation project were positive. An extensive evaluation of 23 ICDPs in Sub-Saharan Africa, Asia and Latin America by Wells *et al.* (1992, p.60) concluded that:

> ...clear-cut lessons for replicating existing projects did not emerge. Instead what stands clearly is that the problems that the ICDPs are attempting to ad-dress are enormous, complex and variable.

Hannah (1992) evaluated ICDPs in sub-Saharan Africa and reported that although the record of addressing entire PAs was better, inadequate fund-ing, particularly in the long-term, had limited project effectiveness in many cases. According to Hannah, even with adequate scope and funding, several design factors emerged as important determinants of project effectiveness. These include the level of technical assistance, the fit of development methods to local conditions, the amount of public support at both local and national levels, the level of enforcement, and the adequacy of project time frame. Yet, there is a clear trend toward increasing numbers of ICDPs in the continent.

In their study of the Usambara Mountains Conservation Project in Tanzania, Stocking and Perkin (1992) questioned the extent to which CWD can be seen as a distinctively different approach to rural development rather than, for example, 'area planning', 'multi-sectoral development', or 'integrated rural development'. Drawing from Livingstone's (1979) analysis of the concept of integrated rural development planning (IRDP), Stocking and Perkin (1992, p.348) argued that the components of CWD are common to IRDP and that:

> ...there *may* be a danger that CWD could degenerate into the methodological limbo of IRDP by failing to address the conservation links.

There is no doubt that the success of protected area management based on the philosophy of CWD and the implementation of ICDPs depends very much on the degree of support and respect awarded to the protected area by neighbouring communities. This is evident from recent research findings on the subject (Newmark et al., 1993; Wells *et al.*, 1992; Hannah, 1992; Brown and Brenes, 1992; Ite, 1996a). For example, an IUCN review of support (buffer) zone projects (Sayer, 1991a) found that success in implementing ICDPs has often been very limited. In addition, the study by Wells *et al.* (1992) revealed very few working models of effective support (buffer) zones. Several factors are responsible for disappointing performance of CWD or ICDPs. Two are particularly significant to the arguments and case study presented in this book.

First, the concept of 'support or buffer zone' has not been adequately defined especially in terms of the objectives, location, area, shape and permitted land uses (Wells *et al.,* 1992). Second, conservation and development agencies tend to offer technological solutions to development problems (Sayer, 1991a, 1991b). As such buffer (support) zone development proposals focus on the application of western industrialised logic to the problems of local people whose culture may be very poorly understood by their distant benefactors. Management efforts in such projects have often failed to adequately consider socio-economic aspects of land use, the patterns of natural resource use and the choices of park neighbours (Schelhas, 1992). Instead, National Park neighbours are perceived as engaging in irrational and environmentally destructive land uses (Schelhas and Shaw, 1992). This stems from the lack of understanding that, although the land uses may be environmentally destructive, they are probably rational given the resources and knowledge available to people engaging in them. Schelhas (1992, p.168) contended that:

> Effective use of buffer zones, biosphere reserves, and other conservation and development strategies to manage land use adjacent to National Park requires tailoring these strategies to the individual situation of the park. This requires a thorough understanding of the biological conservation needs for the park and the socio-economic context in which these programs will be implemented.

There is sufficient evidence to show that conservationists have often neglected socio-economic analysis in the process of protected area planning or, at best, the task is given a very superficial treatment. It is therefore not surprising that cases abound of ICDPs whose work plans are based on little or no socio-economic study of the project area. In spite of the shortcomings of the concept of CWD, there seems to be a general agreement (somewhat paradoxically) that for conservation activities to succeed, ICDPs are essential in forging the new links between local people and protected areas (Wells and Brandon, 1993). The long-term success of the National Park and other PAs requires a shift in management philosophy that combines resource management with a sensitive understanding of the social and economic needs of the local people.

1.5   Synopsis of Chapters

The central argument of this book is that an understanding of smallholder forest farmers and their role in tropical forest loss is vital to attempts to conserve TMF environments in developing countries. This is particularly the case where 'conservation-with-development' (CWD) strategies or 'integrated conservation and development project' (ICDP) approaches constitute the foundation of TMF conservation policy and practice.

The book is organised into nine chapters. The next chapter examines the theoretical approaches of linking small farmers, TMF loss and conservation. Explanations of the causes of TMF loss and the role of small farmers often focus on human ecological and political economic arguments. The human ecological argument is based on the systems approach to the understanding of agrarian change while the political economic debates utilise the structural/historical approach. Within the context of this book, it is argued that decision-making models of agrarian change are also capable of explaining the role of small farmers in TMF loss. Chapter Two therefore underlines the role of political ecology as the best theoretical perspective on the role of small farmers in forest loss, especially at the micro (household) level of analysis.

Chapter 3 focuses on the problem of TMF loss in Nigeria, set against the wider context of deforestation in West Africa. Particular attention is given to the extent, rates and political economy of forest loss in Nigeria, in order to provide the background for understanding the underlying causes and processes of the problem at local level (Chapter 7). To appreciate the philosophy and to enhance the understanding of the reality of conservation in the Okwangwo Division of Cross River National Park (Chapter 8), an overview of the nature and specific problems of environmental and resource conservation in Nigeria is also provided.

Chapter 4 discusses the historical background on the Cross River National Park project, with a particular focus on the Okwangwo Division. It examines the rationale for the establishment of the Division, the theoretical principles and the experience of forest conservation within the Division.

Chapter 5 examines aspects of the physical environment, social and economic development in the Okwangwo Division. It describes the salient physical attributes of the Division including the topography and geomorphology, geology, hydrology and drainage as well as soils. The ecology of the Division, especially the natural vegetation and wildlife resources is also reviewed. From the development perspective, the ethnography, population characteristics, local social and political administration, land tenure system and traditional land use and economy of the Division are examined.

Chapter 6 analyses the nature, pattern and problems of agricultural use of forests within the Okwangwo Division. It discusses the household resource base available for agricultural land use in the villages. Particular attention is paid to farm holdings, sizes, labour force and management as well as capital equipment and farming practices. The prevalent cropping and farm systems, processes of maintenance of soil fertility, the role of household choice of cropping systems in forest loss as well as the major constraints to agricultural production in the study area are examined.

Chapter 7 explores the impact of agricultural land use on the forests in the Okwangwo Division. This is achieved by examining changes in the forest boundary prior to and after the establishment of the Okwangwo Division. New data on forest extent, rates and patterns of loss in the Mbe Mountains complex (a designated conservation and tourism zone of the Division) are presented and discussed. Based on the mapping of forest loss between 1967 and 1993, a spatial and temporal analysis of forest loss in three detailed field study sites is undertaken. The underlying household socio-economic variables and cultural factors responsible for the observed forest loss are examined with a view to articulating the role of small farmers in forest loss in the Division. The implications of the observed changes in forest boundary are discussed.

Chapter 8 evaluates the attempt to conserve the forests of the Okwangwo Division based on the principles of CWD and ICDP discussed in Chapters 1 and 4. It provides a broad overview of the practice of 'conservation-with-development' in the Okwangwo Division. In specific terms, it examines the pilot phase and the European Union-funded stage of the ICDP activities in the Division. The aim is to explore the extent to which the design and implementation of such initiatives can effectively stabilise and check forest loss, especially in areas with a long history of local resource use and management.

Chapter 9 synthesises the major issues raised and highlights the research and policy implications of the major findings presented in the book.

2 Small Farmers, Tropical Forest Loss and Conservation

2.1 Introduction

Explanations of the causes of TMF loss and the role of small farmers often focus on human ecological and political economic arguments. The human ecological argument is based on the systems approach to the understanding of agrarian change while the political economic debates utilise the structural/historical approach. This book argues that decision-making models of agrarian change are also capable of explaining the role of small farmers in TMF loss. This chapter therefore establishes political ecology as the best theoretical perspective on the role of small farmers in forest loss, especially at the micro (household) level of analysis.

2.2 Small Farmers and Tropical Forest Loss: Exploring the Missing Link

The role of small farmers in TMF loss can be examined using the approaches to the understanding of agrarian change suggested by Harriss (1982). These approaches are the systems, structural/historical approaches and decision-making models.

Systems Approach
Studies within the systems approach concentrate on socio-technical systems or on the social systems of agrarian communities, with an emphasis on holistic analysis. While some stress environmental, technological and demographic factors and their inter-relationships within farming systems (e.g. Boserup, 1965), others attempt to make rigorous use of general systems theory to analyse the agricultural sector. Peasant communities are often represented as stable systems, regulated by values in the community, with change in the local society attributed to external factors.

Structural and Historical Approaches
These approaches share attributes with the systems approach described above. Although these approaches have a historical emphasis, they are rooted in

Marxian methodology and examine the ownership and control of resources as well as the associated conflicts. They also consider the exchange and sale of inputs and the marketing of products within the agrarian economy, the 'commoditisation' of production and the incorporation of small-scale producers into markets, among other issues. The structural and historical approaches are concerned with the relationships between the expansion of capitalism and the various forms of production. However, while there is an emphasis on the social character of the 'individual', the individual (in this case the household as an economic unit) is left out of the analyses (Harriss, 1982).

Decision-making Models
Decision-making models of agrarian change focus on individuals and employ techniques of micro-economic analysis. Studies utilising this approach include those concerned with the allocation of resources on the farm and with farmers' responses to markets and innovations. They also include those where individuals are seen as making choices about their values and actions, and thus changing their own societies. Decision-making models provide an excellent understanding and illuminating insights on individuals within the system, but 'the system' itself becomes peripheral to the analysis (Harriss, 1982). By focusing on the 'individuals' and not the 'system', decision-making models are suitable for understanding the economic behaviour of the rural household economy over time and they provide insights into current land use patterns and choices.

2.3 Human Ecological Debates on Tropical Forest Loss

This is the most common explanation of TMF loss, particularly from environmentalists with a strong Malthusian perspective (Hamilton, 1984; Myers, 1984; Whitmore, 1984; Rock, 1996; Laurance, 1999). Advocates of this approach suggest that population growth and pressure create land scarcities which in turn lead to the expansion of agriculture into forested regions (Myers, 1980). Population growth is considered as contributing to forest loss in two main ways.

First, an increase in total national population (rural and urban) increases the demand for food production both from the market system and subsistence farming. In most circumstances, this involves significant land use change particularly with respect to the loss of forest areas to agricultural land use. Several quantitative empirical studies of forest loss tend to support this argument (Brown and Pearce, 1994). For example, Allen and Barnes (1985) analysed the relationship between deforestation and its probable causes in twenty-eight countries in sub-Saharan Africa, Latin America and Asia for the period 1968-1978. They confirmed that deforestation, measured as the rate of loss of forest area, was significantly related to the rate of population growth and indirectly related to agricultural expansion. Their statistical panel analysis also suggested that in the short term, deforestation was due to population growth and agricultural expansion and aggravated over the long term by wood harvesting for fuel and export. Similarly, Rudel (1994) used FAO/UNEP data for 36 countries in Africa, Asia and Latin America to provide empirical support for the significance of population growth for rates of forest loss. His analysis of deforested area for the period 1976-1980 was based on several variables including, forest area, gross national product, and rural population growth. The results suggested that countries with high rates of deforestation are those that have recently experienced rapid population growth.

Second, human ecological arguments maintain that the problem is largely explained and driven by human activities aimed at sustaining an increased local population. From this perspective, it is argued that small farmers are compelled by the Malthusian necessity of a growing population to forgo and or shorten fallow periods. This discourages forest regeneration, leading to gradual reduction in forest extent and cumulative loss of forest area (Rudel, 1994). In other words, population growth in rural areas creates an avenue for the expansion of farmlands while urban population growth leads to the expansion of permanent agriculture as well as deforestation in the immediate vicinity for firewood, building materials and land for settlement. Thus, in both cases, forest loss is the inevitable result of population growth.

The human-ecological framework is limited in its capacity to adequately explain forest loss in a particular location (see also Angelsen, 1999; Angelsen and Kaimowitz, 1999). Bilsborrow and Geores (1994) examined the relationship between demographic factors, land use and the environment on the basis of simple correlation and graphical analyses. They concluded that although there appears to be a positive relationship across countries between rural population growth, land extensification and deforestation, the relationship in each case was weak, and dependent on the inclusion or exclusion of particular outlier countries. The results of their analyses questioned research findings such as those of Allen and Barnes (1985), and Rudel (1994) who found strong relationships between population growth and deforestation based on cross-country data. As Kummer and Sham (1994, p.156) emphasised:

> ... invoking the population pressure argument, without providing the details of how it actually works, does not further our understanding of the deforestation process.

This assertion is supported by several arguments, including the fact that there are at least seven methods available to measure population pressure. These include total population, population density, physiological density, increase in population (in absolute and percentage terms), migration, as well as measures of landlessness or unemployment. While it is not clear which of these measures is the most appropriate for the study of forest loss, there are several inconsistencies concerning their definitions and context of usage. Nevertheless, there is no doubt that population pressure is associated in some way with forest loss.

In sub-Saharan Africa, extensive forest loss in recent years has been evident in Cote d'Ivoire, where the population has increased by 3.6% per annum (Tufuor, 1992). Elsewhere, with high population densities and high population growth rates, the forests in both Kenya and Ethiopia have been drastically reduced. Gabon, the country with the lowest population density and lowest growth rate, has the lowest rate of forest loss. As the World Bank (1992) noted, the falling demand for labour on settled agricultural areas (whether as a result of mechanisation, consolidation of ownership, or economic stagnation) in some countries released a flood of migrants who seek new livelihoods on forest frontiers. Although the best known examples of this process are found in Brazil, Ecuador and Indonesia, the same factors are also evident in West Africa.

Although the approach can explain forest loss at the national or global scale, it provides very limited insight to the local processes (household level) and causes of the problem. The human-ecological argument is deficient in many ways. It fails to incorporate the role of political, historical and socio-economic factors that might influence forest loss especially at the local (household) level. The exclusive use of evidence of population pressure on resources, as the fundamental basis of explanations of TMF loss, therefore contributes little or nothing to advance the current understanding of the spatial dynamics and implications of the problem. There are areas with large populations and relatively low rates of forest loss as well as areas with very few people, a short settlement history and rapid forest loss.

Based on the above, it can be argued that present global generalisations of the relationship between population growth and forest loss have limited value, especially for understanding local (household) processes of forest loss. Furthermore, in some places it would be readily obvious that global demand for timber drives forest loss, while in other geographical locations population growth could be the major factor. As Rudel (1994) noted, in countries with small forest areas (e.g. Burundi, Rwanda and Haiti), a growing population of smallholder farmers are responsible for forest loss while in countries with large expanse of forest (e.g. Brazil), capital investment in frontier regions significantly promote rapid rates of deforestation. Clearly, deforestation has no simple single cause and the role of smallholder farmers varies considerably from place to place through time.

2.4   Political Economic Perspective on Tropical Forest Loss

This perspective on the causes and processes of forest loss builds on neo-classical economics and dependency theory. Neo-classical economic models examine internal factors within a national economy and emphasise the role of public and private capital in tropical forest loss. Proponents argue that faulty incentive systems affect economic and demographic behaviour, especially those centred around the use of common property resources. The role of the state in forest management is seen as central to the political economic argument on the causes of TMF loss (Rock, 1996; Angelsen, 1999; Laurance, 1999). The contention has been that forest loss is encouraged by land tenure rules that confer title to forest lands on parties who 'improve' it by clearing the forest for some other alternative use (see Repetto, 1988).

For example, in Sabah, Malaysia, laws dating from the British colonial period make the state government the holder of all forestry property rights. This vitiates the traditional rights of local communities, but permits any native person to obtain title to forest land by clearing and cultivating it (see also Ite, 1997). The Brazilian authorities tend to settle disputed Amazonian land claims by granting the claimants the title to areas that are multiples of the total forest area cleared for agricultural uses. In Ghana, rights to the use of the resources of the forest that had been governed by local communities were taken over by the central government in the early 1970s. As a result, the rate of forest loss increased because tribal heads no longer had a strong incentive to limit and control shifting cultivation or timber extraction operations. In many other countries, the displacement of traditional communities exercising customary laws over the forest actually weakened controls over resource use (Repetto, 1988).

Dependency theories focus on the external factors which alter production systems and thereby induce environmental decline. Proponents argue that rapid loss of TMF environments coincides with the incorporation of forest regions into an expanding local, national and international economy. In other words, forest loss is accounted for by the penetration of capitalism into the countryside as well as forms of dependency which tie peripheral nations to core states in the global system (Rudel, 1994). Hecht (1985), for example, rejects the simple models of land degradation based on population growth or requirement of export crops. Her analyses suggested that the models of environmental degradation focusing only on the question of production cannot capture the environmental dynamics of speculative economies. The encroachment of the market economy into traditional subsistence societies has been known to lead to the disintegration of traditional cultures which previously exerted the controls needed to ensure sustainable use of common property resources such as the TMF. Changes in socio-cultural controls concerning land use can transform common property resources into open access resources (see for example, Pearce and Turner, 1990). Local management of resources could become difficult, if not impossible, with the result that forest loss accelerates due to lack of effective control by the existing traditional laws and customs.

It is evident that current explanations of the role of small farmers in TMF loss have failed to recognise that smallholder agriculture, agroforestry and pastoral systems can be ecologically sustainable. There is a consensus that most smallholders are rational in decision making and possess adaptive behaviour and skills based on their understanding of the local environment. In addition, arguments about the destructive influence of small farmers often underestimate the role of external factors such as government policies, class position and land tenure, in the internal (household) organisation of resource management. Yet these issues play significant roles in influencing household decision-making concerning land use and the management of the forest environment.

2.5   Towards an Alternative Perspective

The third approach for establishing links between TMF loss and smallholder agriculture can be derived from the decision-making model of agrarian change (Ite, 1996b). The loss of TMF results from cumulative land-use decisions. To understand the role of small farmers in such losses, it is important to examine their decision-making process concerning the use of land. Blaikie (1985, 1989) developed an approach to the general issue of environmental degradation, which is relevant to the arguments in this book.

According to Blaikie, individual decision-making units (e.g. households) are conceived as choosing a form or forms of income generation to fulfil some objective function at each time period, and usually each season of the year. These objective functions vary according to the social, political and economic circumstances of the household, and affect the ways in which the household views the array of income opportunities. Each household has at its disposal both physical and human resources e.g. land, labour, equipment, education, ethnicity and extended families. These resources are used to determine the income opportunities open to each household. Such opportunities may be within the farming system, involving cropping rotations, for example. Alternatively, they could be non-agricultural activities (e.g. trade and remittances, wage labour, artisanal production), with no significant implications for the deterioration or loss of environmental quality.

Each of the income opportunities has an 'access qualification'. Income opportunities with the least severe qualifications tend to be over-subscribed (e.g. wage labour), while those with the highest qualifications (usually capital) attract the highest reward. However, each income opportunity can only be taken if the household has access to the necessary qualifications (labour, land, capital and equipment) and the social networks necessary to mobilise them (Blaikie, 1989). At any period of time, households possess a range of assets whether purchased or inherited. They can also buy or sell assets, thus affecting their ability to choose income opportunities. Assets could be a 'windfall' (e.g. nearness to market) or 'inherent' (e.g. ethnicity), and all change through time with the waxing and waning of household fortunes (Blaikie, 1985).

Tropical forests provide households in developing countries with an array of income opportunities, including farming, hunting and gathering of non-timber forest products (see Byron and Arnold, 1999). However, these opportunities may not be open to many households because of several social, political and economic constraints. Small farmers therefore play a fundamental role in TMF loss through the deployment of resources and the resource allocation techniques used in the process. This is further supported by households' response to both economic incentives and innovations prevailing in the wider political economy. In linking small farmers to TMF loss, it is therefore important to note that farming in TMF environments involves periodic clearance of vegetation, cropping regimes and abandonment of the plot with the loss of soil fertility. These have implications for the rates and extent of deforestation.

However, of particular significance are the factors influencing the patterns of clearance of forest vegetation. According to Rudel and Horowitz (1993), the size and shape of forest influences how people clear land. Where small forests predominate, people clear land along the forest fringes while in areas with large forests; land is cleared along the fringes as well as along corridor-shaped tracts towards the centre of the forest. The people who clear the land and the events that spur them into action vary considerably from fringe to corridor clearing. Clearance along the forest fringes and subsequent colonisation occur in many ways. Farmers could advance along a broad front into a forest. For example, according to Dourojeanni (1979, cited in Myers, 1984, p.150), in the Amazon:

The population overflowing from the Andes down to the Amazon plains do not settle here. They advance like a slow burning fire, concentrating along a narrow margin between the land they are destroying and are about to leave behind, and the forests lying ahead of them.

As agricultural expansion therefore pushes the farm frontiers into the forest, the assumption is that farmers would prefer to clear the forest area nearest to the village in the first instance. However, population growth makes it imperative for farmers to proceed further from the village to clear fresh forest areas. This could result in the relocation of entire villages as distances increase to the newly cleared areas. In some regions (e.g. Latin America), decline in soil fertility levels motivates farmers to cede or sell the land to cattle ranchers and proceed farther to new forest areas where the cycle of clearing and abandonment continues (Rudel and Horowitz, 1993).

The relevance of small farmers' decision making in contributing to forest loss can also be examined from a socio-economic perspective. For example, in agro-ecological terms, southern Nigeria is a zone of tree (e.g. cocoa, citrus, banana) and root (e.g. cassava, yams, cocoyams) crops. The bulk of agricultural production occurs under traditional systems characterised, for example, by the use of non-mechanical power and small holding size due to land fragmentation (Abumere, 1983; Nwafor, 1982). There is growing evidence that in response to the need to improve the economic well-being and rising social expectations of their households, small farmers including those in southern Nigeria, have made substantial efforts towards the better management of TMF environments as farms (NEST, 1991). These farmers are influenced by several socio-economic factors most of which are external to households, especially the land tenure system. Despite the limitations of socio-cultural factors, small farmers in Nigeria respond to change by balancing wider traditional expectations, individual household welfare and the economic viability of change. The point is that given a favourable setting, small farmers in Nigeria are capable of being responsive to ideas of higher income and introduce profitable new crops as well as adopt new farming practices (see Williams, 1978; Helleiner, 1966). A good case in point is the cultivation of cocoa as an export crop by farmers in northern Cross River State (see also Chapter 6), a region noted for the cultivation of food crops such as yams.

Small farmers view TMF environments as the source of livelihood for the present and future generations. There is evidence that small farmers in TMF environments are indeed conservationists (Clay, 1988; Ite, 1997). They do not cause permanent degradation of soils either deliberately, except in extreme circumstances, or accidentally. For example, although shifting cultivation is a widespread agricultural system, it has not generally resulted in long-term elimination of forests (Myers, 1980; Coomes, *et al.*, 2000). There-fore, in designing TMF conservation projects and policies, there is a need to build on the knowledge of small farmers concerning their environment and their practices in relation to the sustainable use of the environment (see also Fujisaka, 1989). Such knowledge can only be garnered through a study of individual farming units (households) as the focus of the analysis. This is the approach adopted in Part III (i.e. Chapters 5-8) of this book.

2.6 Political Ecology Approach as the Missing Link

Since the early 1980s, the political ecology approach has been considered as a fruitful research agenda in Third World studies. Although there are different formulations and interpretations of political ecology, all are drawn from the wider political economy perspective (see for example, Bryant, 1992; Brown, 1998; Kalipeni and Oppong, 1998). Political ecology offers the framework for understanding the political causes, conditions and ramifications of envi-ronmental change. As a critical approach, it focuses on the political, eco-nomic, historical, social structures and processes that underlie human prac-tices leading to degradation (Bryant, 1992; Neumann, 1992). In other words, it is a model that specifically focuses on the linkages between society, politi-cal economy and the environment. It provides an excellent framework for examining the relationships between local patterns of land use and the larger political economy (cf Blaikie and Brookfield, 1987; Bassett, 1988; Stonich, 1989). The analysis of land degradation issues against the background of po-litical ecology embodies a number of common elements, three of which are significant to the arguments presented in this book.

Firstly, political ecology requires a 'bottom-up' analytical approach, starting from the smallest decision-making unit concerning the use of land and extending upwards to a larger unit (Blaikie, 1985). There is a micro-focus on land users or managers (e.g. peasants, state forestry departments) who may have different decision-making environments as well as claims or demands upon the same tract of land. Thus land users (and their existence in a web of social relationships which link them to wider political economic structures extending beyond the locale of land degradation) constitute the central analytical issues to be addressed (Neumann, 1992). Although the house-hold is the smallest unit where land-use decisions are made, there are inher-ent social, economic and political inequalities within this unit, especially in relation to age and gender. This could manifest itself, for example, in the allocation of privately controlled resources and labour for land-use, with sig-nificant implications for land-use and conservation (see, for example, Leach,

1994; Davies and Richards, 1991).

Secondly, political ecology is scale-dependent. The analysis requires 'a chain of explanation' in which the linkages of the local (in this case, households) relationships with the wider (i.e. regional, national, global) geographical and social setting are traced (Blaikie and Brookfield, 1987). The approach is to examine the ways in which the land users' behaviour is influenced by social structures and processes within and outside their immediate environment (Neumann, 1992).

Thirdly, for an understanding of the development of social relationship and their links to land degradation, the political ecology perspective places emphasis on historical analysis. It examines the transformation of indigenous systems of resource management in the process of incorporation into the global economy. The historical analysis of a region's linkage to the global political economy can provide a much better understanding of the effects and impact on land use and conservation practices, of current changes external to a local setting (see Blaikie, 1985; Stonich, 1989).

It is clear that 'human-ecological' and 'political-economic' arguments on the causes and processes of TMF loss provide an inadequate basis for understanding the problem of TMF loss and its socio-economic implications at the household level, or the linkages between environment and economy at different spatial scales (Ite, 1996b). Both perspectives tend to over-simplify the complex reality of the actual processes of TMF loss. There are several intervening variables not captured by such analyses. If protected forest areas and associated conservation programmes are to yield expected results, there is a need for a better and in-depth understanding of the complexity of the links between forest loss and forest farm economies. This requires an analytical approach with the smallholder farmer (land manager) as the focal point of the study (see Ite, 1995).

The problem of human causation of land degradation is very complex (Blaikie and Brookfield, 1987). It occurs in a variety of social and ecological circumstances. It is clearly futile to search for uni-causal models of explanation, especially since there are several conjectural factors operating at one place and time. The political ecology approach to explaining environmental change can handle the complexity of the process of TMF loss. The approach is therefore be the most suitable conceptual basis for the explanation and analysis of both the roles of small farmers in TMF loss and their involvement in conservation policy.

2.7   Conclusion

The causes, rate and location of farmer encroachment on the forest cannot be understood without considering decision-making in farming households against the wider context of the economy and certainly not in terms of a simple neo-Malthusian and population growth model. This chapter therefore acknowledges the superiority of the political ecology framework since it incorporates a 'bottom-up' analytical approach.

As the case study in Part III will show, if conservationists misunderstand the dynamics of the encroachment on the protected forest boundary by small farmers, responses to such encroachment will be ineffective. Furthermore, if such responses are 'top-down' or badly organised, these could exacerbate the problem of forest loss. This is where the political ecology approach becomes very useful in understanding the contextual causes of forest loss and how local forest-dependent communities could be effectively integrated into forest conservation programme.

PART II: ENVIRONMENTAL MANAGEMENT ISSUES IN NIGERIA

3 Tropical Forest Loss and Resource Conservation in Nigeria

3.1 Introduction

This chapter focuses on the problem of TMF loss in Nigeria. It is set against the wider context of deforestation in West Africa. There is particular emphasis on the extent, rates and political economy of tropical forest loss in the Nigerian context. The view is to provide the background for explanation and analysis of the underlying causes and processes of the problem at the local level (Chapter 7). To appreciate the philosophy and reality of forest conservation in the Okwangwo Division of the Cross River National Park (Chapter 8), the nature and specific problems of environmental and resource conservation in Nigeria are highlighted.

3.2 Regional Context of Tropical Forest Loss: West Africa

Using the 1959 vegetation map of Africa produced by the *Association pour l'Etude Taxanomique de la Flore d'Afrique Tropicale*, Sommer (1976) calculated that TMFs in Africa (including those in Madagascar, Mauritius and Reunion) originally covered 3.63 million km^2. One-fifth (19%) of the total TMF for Africa was in the West African sub-region. However, by the beginning of last century, only 28% of the West African total remained. This suggested a decrease of 72%, which was far above the global average of 41.2% (Martin, 1991). In spite of the controversy that greeted Sommer's work and doubts concerning its reliability, the results provided the first clear message of the rapid rate of forest loss in the West African sub-region.

Subsequent estimates of the annual rate of tropical deforestation in the West African sub-region have varied (Nectoux and Dudley, 1987; Martin, 1991; Tufuor, 1992; Fairhead and Leach, 1998). Nevertheless, all suggest that there has been a general and significant decrease in the TMF extent in West Africa in recent decades. Regarding the possible explanations for the observed losses of TMF, Martin (1991) argued that West Africa's proximity to Europe influenced its trade relations for an appreciable period of time, thus

29

causing rapid forest loss. In Ghana and Nigeria, TMF areas were increasingly cleared for export crop plantations (e.g. cocoa and oil palm) during the last quarter of the 19th century. The introduction of mechanised logging operations also resulted in rapid forest loss in the sub-region during the early 20th century.

Until recently, the 1980 FAO estimates of forest extent in West Africa (Table 3.1) was widely cited as probably the best estimate of remaining TMF and their distribution in the sub-region. However, unlike other estimates such as those of Lanly (1982) and Myers (1980), the FAO estimates were derived from a uniform date, and provided information essentially for timber utilisation. It placed emphasis on the assessment of open and closed forest formations as well as undisturbed areas and forest areas subject to exploitation. Despite its shortcomings, the FAO estimates confirmed a decrease in forest extent and the rapid rate at which new areas were being opened mainly due to timber exploitation (Martin, 1991). There is no doubt that that in some countries (e.g. Liberia), it is possible that there are substantial areas of TMF left, while in others (e.g. Benin), only a small proportion of the original TMF cover can be found.

The rate of deforestation varies widely among the countries in West Africa. In 1980, the closed moist forest area of Sierra Leone, Liberia, Cote d'Ivoire, Ghana, Nigeria and Cameroon covered a total of 35.69million hectares, with an annual rate of deforestation of 0.75million hectares (2.09%). In Cote d'Ivoire alone, the annual deforestation rate was estimated at about 300,000 hectares (or nearly 7%), in a country where forest is reported to have once covered 45% of the total land area (Tufuor, 1992). In 1900, there were 16.4million hectares of tropical forest in Cote d'Ivoire. By the end of 1981 only 4.5million hectares was left. Tufuor (1992) estimated that if these rates of deforestation were sustained without afforestation, the remaining forest would have a life expectancy of 8-9 years.

Different causes of TMF loss are important in different places. For instance, in the 1980s, TMF loss in Cote d'Ivoire was propelled by commercial logging, an intensive drive for cash crops, and the activities of farmers who cut down trees without replanting them (Bourke, 1987). In early 1983, the country lost (permanently) about 10% of its already depleted forests to bush fires (*West Africa*, 4 April 1983, p.821).

In a study of six African countries (Cameroon, Congo, Cote d'Ivoire, Gabon, Ghana and Liberia), Poore *et al.* (1989) observed that there was hardly any genuine sustained-yield forest management whether extensively or intensively organised. Past efforts to replant in the West African sub-region amounted to only 0.314million hectares; one hectare of forest planted for every 24 hectares cleared. As Tufuor (1992) observed, tropical Africa as a

Table 3.1   Original and Remaining Tropical Moist Forest in West Africa

Country	Approximate original extent, (km²)	Remaining extent (based on FAO 1980 data) (km²)	Extent from maps (km²), with dates	% Remaining (based on FAO 1980 data)	% Remaining (based on map data)
Senegal	27,770	2,200	2,045 (1985)	7.9	7.4
Gambia	4,100	650	497 (1985)	15.6	12.1
Guinea Bissau	36,100	6,660	-	18.4	-
Guinea	185,800	20,500	7,655 (1989)	11.0	4.1
Sierra Leone	71,700	7,400	5,064 (1989-90)	10.3	7.1
Liberia	96,000	20,000	41,232 (1989-90)	20.8	43.0
Cote d'Ivoire	229,400	44,586	27,464 (1989-90)	19.4	12.0
Ghana	145,000	17,180	15,842 (1989-90)	11.8	10.9
Togo	18,000	30,040	1,360 (1989-90)	16.9	7.6
Benin	16,000	470	424 (1979, 1989-90)	2.8	2.5
Nigeria	421,000	59,500	38,620 (1989-90)	14.1	9.2

Source: Adapted from WCMC (1992).

whole had a replacement rate of only 1:10, compared with 1:4 for tropical Asia. The extraction of timber was accompanied by little or no reinvestment of the derived benefits. The reasons for this failure to achieve sustainable forestry practices has been attributed to the weakness of forest management institutions (with respect to land use, forest policy and legislation) and the lack of information and data required for sustainable forest management.

3.3   Tropical Forest Loss in Nigeria: Extent, Rates and Proximate Causes

Vegetation untouched by human activity probably no longer exists in Nigeria (Wood, 1993). Writing in 1949, Keay (cited by Galletti *et al.*, 1956, p.19) observed that:

> Nigeria has a population of 23,000,000, and a population map of Africa shows clearly that, apart from the Nile Valley and urban centres such as Johannesburg and Cape Town, it is the most densely populated part of Africa. There is good reason to believe that it has been relatively inhabited for at least several centuries. This relatively dense population has of course had effects on the vegetation, so much so that apart from swamps and mountainous regions it is true to say that there is to-day little or no virgin forest in Nigeria. Much of the

vegetation can still be regarded as natural but even in the finest Nigerian forests (e.g. in Benin) one is constantly finding evidence (ecological and archaeological) of past farming or occupation by man.

The area of primary uncleared TMF has now been reduced to small pockets in the Niger Delta, the west (e.g. Ondo and Ogun States) and southeast (Cross River State) Nigeria. Most of the remaining areas of TMF are within designated protected areas (i.e. Forest Reserves or National Park). Data presented in Table 3.1 suggest that the approximate original extent of TMF in Nigeria was 421,000km^2. However, at the end of 1980, only 14% (or 59,500km^2) of the total extent was left. From 1980 to 1990, Nigeria's TMF shrank from 14% to 9.2% (i.e. 38,620 km^2), which was equivalent to 4.2% of the total land area in Nigeria.

As Ite (1995) noted, this decrease can be seen as a reflection and function of different deforestation measurement methods and may not be a real change in forest extent. However, there is no doubt that rate of TMF loss in Nigeria has been high, although estimates vary widely. For example, the Nigerian Environmental Study/Action Team (NEST, 1991) asserted that Nigeria was losing up to 350,000 ha of forest and natural vegetation annually. Using the FAO/UNEP 1980 forest inventory data and data on the growth of population and gross national product, Umeh and Harou (1992) suggested that the annual deforestation rate in Nigeria was 3.5% (see also Schreiber, 1993). Generally, estimates of the rate of forest loss in Nigeria are limited by the lack of any extensive resource surveys in the country (NEST, 1991). Although data are poor, it is now clear that the process of TMF loss in Nigeria is particularly well advanced with probably not more than 10% of the original TMF cover remaining in the country. Insights from existing studies suggest that at the national level, human ecological factors and political economic processes influence forest loss. The human-ecological factors relate to population growth and land pressure as well as woodcutting for fuel. The political-economic processes include forest management policies, patterns of land tenure, levels of infrastructural development, urban expansion and industrialisation (Jagannathan *et al.*, 1990; Caldecott and Morakinyo, 1996).

Forest management in Nigeria is the statutory responsibility of forestry departments in the various states of the federation. However, the main concern of such departments has not always been environmental in orientation (Blench *et al.*, 1999). As such, the problem of forest loss was made worse by management policies adopted to reflect the wider context of the Nigerian economy. For example, in the 1980s many state forestry departments concentrated on the establishment of industrial plantations for gmelina (*Gmelina arborea*) and teak (*Tectona grandis*), which involved clear felling of natural forest (Umeh, 1992; Osemeobo, 1990). During the period under review, Cross River State in southeast Nigeria accounted for 9% of the total area of all plantations and 17% of the area of *gmelina* plantations in the country (Umeh, 1992).

The second form of forest management by state forestry departments was the introduction of *taungya* farms in some forest reserves as a tree plant-ing policy (Osemeobo, 1988; Osemeobo, 1993a). This served as a protective belt around accessible parts of forest reserves, protecting the natural forest behind from encroachment (Umeh, 1992). Under *taungya*, an area of forest was allocated to local farmers to be cleared after it had been logged. In ex-change for farming rights, the farmers were required to plant seedlings of useful tree species including teak and *gmelina* provided by state forestry de-partments. In the long term, these trees were intended to eventually replace the food crops (Oates, 1995).

The loss of TMF in Nigeria was also accelerated by the ban on the importation of timber in the late 1970s and early 1980s. Increased demand for wood products within Nigeria served as the catalyst for extensive logging within and outside forest reserves. Natural regeneration techniques were aban-doned because they could not produce enough timber to satisfy internal de-mand nor export (Morakinyo, 1992; Umeh, 1992; Osemeobo, 1990). Never-theless, it is important to note that the rate of exploitation of the forests in Cross River State in southeast Nigeria was relatively low compared with other states in the country. This was attributed to the rugged terrain and to the fact that until 1973 there were no bridges over the Cross River to allow the easy movement of timber to other parts of Nigeria (Lowe, 1992).

Agriculture has been a significant cause of TMF loss in Nigeria. At the dawn of Nigeria's political independence in the 1960s, agriculture was the mainstay and driving force of the national economy. It provided employment for almost 80% of the nation's labour force (Okurume, 1993), and small-holder farmers produced enough food to feed the entire population (Ekpo, 1993; Forrest, 1993). The economy of the rainforest belt was transformed by cash cropping. Extensive and rapid loss of forest outside the Forest Reserve areas have been reported (Morakinyo, 1992; Osemeobo, 1993a; Osemeobo, 1990 and 1988). In Bendel, Cross River, Ondo and Oyo states, large areas of forest were cleared and replaced with plantations of cash crops such as cocoa, rubber and coffee. Other areas were cleared and devoted to food crop produc-tion.

The theoretical link generally drawn between population growth and forest loss has been well established in Chapter 2. Historical records suggest that since pre-colonial times the population density in southern Nigeria was relatively high compared to the rest of Africa. During this period, significant areas of forest were lost especially in areas around the ancient kingdoms of Ife, Oyo and Benin in the west and Iboland in eastern Nigeria. The decline of these kingdoms as a result of wars, disease and slave trade led to the regeneration of forest which could scarcely be distinguished from primary forest (Morakinyo, 1992).

High population densities in the rural areas, particularly in southeast Nigeria, have resulted in land hunger, farm fragmentation and the shortage of fuelwood. The effects include widespread deforestation, decimation of wildlife, increased environmental aridity and desertification in the dry belt (NEST, 1991). Reduced fallow periods on farmland have resulted in soil impoverishment, declining yields, poor harvest and soil erosion. For example, Lal and Okigbo (1990) reported that fallow periods in Imo State had declined from 1-9 years to 0-6 years - a period much shorter than the 5-7 years required for restoring soil fertility. Consequently, there has been considerable and steady out-migration of farmers from such areas to settle and farm in sparsely settled areas of the country with sufficient and perhaps, better agricultural land (NEST, 1991; Udo, 1993). National population growth motivated government dereservation of approximately 56,000 hectares of gazetted forests for agriculture (NEST, 1991; Morakinyo, 1992). The extent of such dereservations in southern Nigeria favoured the expansion of food crop production by small farmers as well as plantation development by capitalist farmers and parastatals (Osemeobo, 1988).

3.4   Environmental and Resource Management in Nigeria

The history of environmental management in Nigeria since 1970 has been catastrophic, as reflected in the state of the forests, which have been greatly reduced in area and condition. Until recently, environmental conservation was considered as a 'middle-class' cause, as it has been mainly promoted by middle-class intellectuals and their wealthy sympathisers (Caldecott and Morakinyo, 1996). Nonetheless, there has been a shift in conservation thinking and development in Nigeria. From a largely governmental affair, it has become a shared concern between the government, non-governmental organisations (NGOs) and the people (NEST, 1991; Blench *et al.*, 1999). There are several pieces of domestic legislation on natural resource conservation. These include the various federal laws as well as state and local government edicts relating to soil, land use, mining laws, water, plant, animal, and fisheries resources. The concern of non-governmental organisations is evident in the number of such organisations actively involved in conservation and related activities (Table 3.2). For example, the Nigerian Conservation Foundation,

Table 3.2   Selected National Conservation Organisations in Nigeria

Organisation and Year Established	*Main Activities*
Nigerian Field Society (1930)	• Meetings and lectures on conservation and related subjects. • Documents and popularises the natural history of Nigeria.
Nigeria Conservation Foundation (1982)	• Attracting attention and funds from international conservation organisations, government, and private donors for supporting conservation projects.
Nigerian Environmental Study/Action Team (1987)	• Raising awareness and influencing public policy for better management of the environment.

an affiliate of the World Wide Fund for Nature (WWF) was established in 1982 and has proved to be a very vocal lobby for species and habitat conservation.

Several international environmental organisations, bilateral development institutions and aid agencies have also taken keen interest in environmental conservation in Nigeria. For instance, the UK Department for International Development has been involved in the sustainable management of the TMF resources in Nigeria, especially in Cross River State. The European Community and WWF-UK contributed significantly to the investments in the Cross River National Park Project (see Chapter 5). Nigeria is party to various international treaties and conventions concerning conservation including the African Convention for the Conservation of Nature and Natural Resources.

Significant measures were also undertaken for in-situ conservation of plant and animal biodiversity in Nigeria at the local, state and national levels of governance. These include the establishment of National Parks and Game Reserves. In addition, several ex-situ conservation facilities including zoological gardens and gardens have been provided (FEPA, 1992). Prior to May 1999 when the Federal Ministry of Environment was created, there were other notable national efforts towards environmental and resource conservation in Nigeria. These include the formulation in 1986 and adoption in 1988 of a National Conservation Strategy (NCS) and a National Conservation Education Strategy (NCES). The NCS emphasised the need for inter-sectoral analysis of natural resources issues whenever investment decisions were contemplated.

Various aspects of the NCES were implemented by the Nigerian Conservation Foundation, especially in the Cross River National Park Project (Chapter 5). The promulgation of the Endangered Species (Control of International Trade and Traffic) Decree No. 11 of 1985 was another significant milestone in Nigeria's conservation efforts. The decree provided the necessary legal backing for dealing with offences under most international conventions, especially the Convention on International Trade in Endangered Wild Species of Fauna and Flora (CITES).

In 1989, the Natural Resources Conservation Council (NARESCON) was established as the apex organisation for natural resources conservation in Nigeria. In accordance with the provisions of Decree No. 50 of 1989, NARESCON had the responsibility of co-ordinating various ideas, policies and strategies into a national focus for the conservation of renewable natural resources. It formulated a Conservation Action Plan for Nigeria, which served as a companion to the NCS (Federal Republic of Nigeria, 1991a; NARESCON, 1992). NARESCON played a decisive role in the events leading to the establishment of the Federal National Parks Service (FNPS). Until May 1999, the FNPS was a division within the Federal Ministry of Agriculture and Natural Resources.

The Federal Environmental Protection Agency (FEPA) was concerned with the implementation of the National Environmental Action Plan. FEPA channelled funds to the various state environmental protection agencies, in accordance with the principle of devolving most aspects of environmental management, including forest management, to state governments. The role of the Federal Government has been mainly to provide advice and co-ordination for state government activities, to provide law enforcement, to undertake research, as well as funding and implementation of certain projects (Stuart *et al.*, 1990; Blench *et al.*, 1999). The Federal Ministry of Environment has now taken over the responsibilities of both FEPA and NARESCON.

3.5   Protected Area System and National Parks in Nigeria

Conservation areas constitute at least 5% of the total land area in Nigeria (FEPA, 1992). Protected areas come under several protection systems including Forest Reserves, Game Reserves, and National Parks. The establishment in 1899 of the first Forest Reserve in Nigeria marked the beginning of formal action to conserve Nigeria's natural vegetation. The original policy was to conserve 25% of the total land area (Lowe, 1984). However, the approximately 10% of land held under Forest Reserves fell far short of this target. According to NEST (1991), this compared very poorly with those of some industrialised countries such as France (25%) and Japan (67%). Under the Forest Reserve system, land is held in trust for the indigenous people and government controls resource exploitation in the area. The constitution of Forest Reserves involved elaborate procedures for consultation between government and all interested parties, including the different arms of government as well as local communities. All affected persons were allowed to voice objections or to request continuing rights and privileges within the reserve, such as hunting or fishing, collection of wild fruits or materials, or rights of way, before legal constitution and consolidation of the reserve. Compromises were reached with them, and continuing rights and privileges were normally granted to local communities to the exclusion of outsiders (Lowe, 1984, 1992). In forestry circles, emphasis was laid on Forest Reserves partly because they have been the most effective means of conserving natural vegetation. Forest

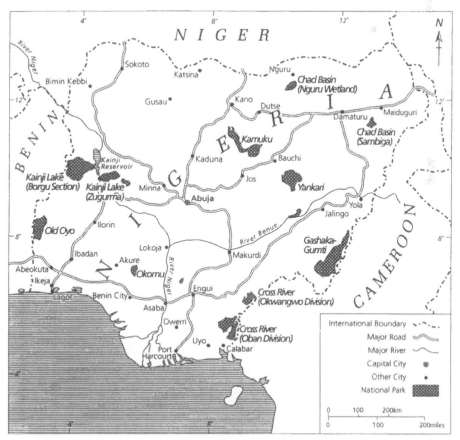

Figure 3.1   National Parks in Nigeria

Reserves have remained the responsibility of State Governments who in most cases managed them primarily for timber extraction (see Blench *et al.*, 1999). Although they enjoy some form of protection under Forest Law, the conservation status of reserves is generally weaker than the National Park system as they are usually poorly protected from encroachment.

Based on the IUCN (1994a) categorisation of protected areas, there are 8 National Parks (IUCN protected areas category II) in Nigeria (see Table 3.3 and Figure 3.1). The main objective of the National Park system in Nigeria is to protect the natural heritage, and this is the responsibility of the Federal Government. Each National Park has a management board whose functions include ensuring the implementation of the National Parks policy of the Federal Government of Nigeria. The activities of the National Parks are co-ordinated by the Federal National Parks Service (FNPS), a division of the Federal Ministry of Environment. All the National Parks in Nigeria are federal institutions, and tend to be seen as federal 'projects' in the various states where they are located (Table 3.3). They serve as useful vehicles for attracting federal and donor funds, but nevertheless as institutions not under state government control. State governments do not provide financial support, their contribution having been mainly the initial grant of land for the parks.

The Strict Nature Reserves in Nigeria are mostly relatively small and are intended to conserve various examples of primary vegetation. They are within Forest Reserves but are not protected by any specific legislation. Thus they have no legal status to distinguish them from Forest or Game Reserves. Game Reserves are controlled by the states, and incorporate areas where hunting is supposed to be strictly regulated. They are areas to be set aside for the

Table 3.3   National Parks in Nigeria

National Park	Location (State)	Area (ha)	Year established
Chad Basin	Borno, Yobe	230,000	1991
Cross River	Cross River	400,000	1991
Gashaka-Gumti	Adamawa	586,000	1991
Kainji Lake	Niger	532,000	1975
Old Oyo	Oyo State	253,000	1991
Yankari	Bauchi State	225,000	1991
Okomu	Edo	n/a	1999
Kamuku	Kaduna	n/a	1999

Source: Adapted from IUCN (1994b) and updated.

conservation, management and propagation of wildlife and the protection of their habitat (Lowe, 1992; WCMC, 1988).

Wildlife conservation in Nigeria began with the promulgation of the Wild Animals Preservation Ordinance in 1916. Between 1916 and 1960 when the country gained independence from British colonial rule, the number of Game Reserves rose from 12 to 29 (Anadu, 1987). However, very few concrete efforts were made to protect game from over-hunting largely due to official apathy during colonial times. There was low priority rating of wildlife as reflected in inadequate funding and administrative arrangements, together with weak enforcement and inadequacies in existing wildlife laws. Other factors include excessive demand for land, bushmeat and fuelwood by a rapidly growing human population, as well as a traditional lack of concern for the welfare of wild animals (Anadu, 1987).

3.6 Challenges of Environmental and Resource Conservation in Nigeria

The conservation of environmental and resources in Nigeria has been beset by several challenges (Ite, 1998; Blench *et al.*, 1999), four of which are particularly significant. First, the policy and legal instruments are poor. Except perhaps for the forestry sector for which legislation was enacted in the colonial period, environmental and resource conservation efforts in Nigeria have largely been fragmented, uncoordinated and without clearly defined overall goals related to national policy. They have been characterised by piecemeal strategies and *ad hoc* approaches. Several pieces of legislation directed at specific resource problems were enacted without proper research. As a result they have not been sufficiently far-reaching or comprehensive to resolve their targeted issues or problems.

For example, the Land Use Decree of 1978 vested authority over all land in the country in government in trust for the people and to be used for the common benefits of all Nigerians (NARESCON, 1992). The decree developed rules about land tenure in Northern Nigeria (with a very different set of traditional land tenure rules, agricultural ecology and economy) and imposed on the rest of the country (see also Chapter 5). The decree was deeply resented in southern Nigeria where large-scale evasion is clearly evident (Udo, 1990). Another example of policy lapses in Nigerian conservation efforts relates to the Endangered Species (Control of International Trade and Traffic) Decree No. 11 of 1985. The decree offered total protection of fairly common species (e.g. kites) and permitted trade, under licence in endangered species such as cranes, secretary birds, and ostriches. It did not offer any protection to the country's amphibians, nor did it cover plants, even though many are threatened.

Second, there is lack of data for policy formulation and project planning in Nigeria. Rational use of resources depends on accurate and complete data. As NARESCON (1992) admitted, lacking their own programmes for conservation and vital data and information about resources, successive Nigerian governments relied heavily on ideas, concepts and programmes suggested or recommended by consultancy firms, local pressure groups and international agencies. In addition, there has been great dependence on special commissions and *ad hoc* task forces in examining specific resource and environmental problems. A particularly good example is the suggestion to the Nigerian government in April 1988 by WWF-UK for the establishment of the Cross River National Park project. It has been argued that since the idea for the establishment of the Park project did not originate from within the country, the political will and perhaps the logistic capacity to implement the project successfully remains doubtful (Ite, 1998). This is based on the huge financial implications and the competing interests in the resources of the National Park area.

Third, the enforcement of existing laws and regulations on environmental resources is fraught with problems, resulting in the practice being poles apart from the theory. Most laws in Nigeria exist in the statute books but are not enforced in practice. For example, cases abound of the breach of laws concerning all areas of natural resource conservation, including those relating to wildlife, fishing and forestry.

Fourth, institutional problems beset environmental and resource conservation initiatives in Nigeria. None of the natural resource sectors in the country has a well-integrated management and development structure from the village to the Federal Government level. Accordingly, co-ordination and co-operation between the different levels of governance becomes difficult. In many cases, there has been duplication, and even conflict between programmes (NARESCON, 1992; Holland *et al.*, 1989).

3.7 Conclusion

The loss of TMF in the West African sub-region has reached an advanced stage due to the interaction of human ecological and political economic factors over time. Although there is a general decrease in the forest extent, it has been acknowledged that estimates of annual rates of loss vary depending on the source. There are also variations between countries in the importance of the factors influencing forest loss. In Nigeria, there is a relationship between economic development and TMF loss. The expansion of the agricultural land area and population growth constitute the key factors in TMF loss in Nigeria. An understanding of environmental and resource conservation in Nigeria is important for appreciating the challenges of forest conservation, discussed in Part III of this book.

4 The Cross River National Park Project

4.1 Introduction

The Cross River National Park is located in Cross River State, southeast Nigeria (Figure 4.1). The Park, named after the Cross River (which also lends its name to the State), covers an area of approximately 4,000km^2. It consists of two divisions: Oban in the south and Okwangwo in the north (Figure 4.1). The Oban Division was named after the Oban Hills, which dominate the topography of the area. With an area of approximately 3,000km^2, the Oban Division was carved out of the Oban Group Forest Reserve.

This chapter discusses the historical background on the Cross River National Park Project, with a particular focus on the Okwangwo Division. It examines the rationale for the establishment of the Division as well as the theoretical principles and practical realities of forest conservation in the Division.

4.2 The Cross River National Park Project: Historical Review

The origin of the Cross River National Park Project (CRNPP) dates back to the early 1930s when the Oban Group and Boshi/Okwangwo Forest Reserves were established (Ebin, 1992). Proposals to conserve the unique wildlife resources of the area were made repeatedly thereafter, although no actions were taken until the late 1980s (Caldecott *et al.*, 1989, 1990a). Significant efforts were made towards the implementation of the conservation proposal at that time. These efforts were influenced by several publications (e.g. IUCN, 1986; 1987), and reports from other independent studies that identified these forests as requiring special conservation attention. Following the constitution of the adjacent Korup forest in Cameroon into a National Park in 1987, action began in 1988 in Nigeria to establish the CRNPP. The principal interest groups in this process were the Federal Government of Nigeria (FGN), the Cross River State Government (CRSG), WWF-UK, and the Nigerian Conservation Foundation (NCF). According to Caldecott *et al.* (1990a), WWF-UK proposed the concept of an Oban National Park to both the FGN and CRSG. The latter approved in principle the constitution of a National Park in the Oban

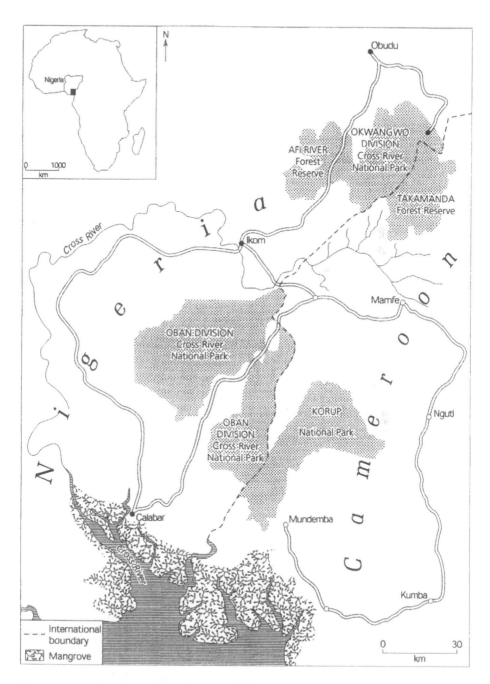

Figure 4.1   Cross River National Park

and Boshi-Okwangwo areas. This was followed by joint preparation by WWF-UK and NCF of preliminary descriptions of the Oban Park area, and the accreditation of a Project Manager to the CRSG by WWF-UK. By late 1988, the NCF began the management of a Gorilla Conservation Project initiated in 1987 in the Mbe Mountains area (see Figures 4.2 and 4.3) with a field research station at Kanyang 1 village.

In early 1989, HRH Prince Philip, the Duke of Edinburgh visited Cross River State to launch the CRNPP in his capacity as the International President of WWF. He signed a technical and financial agreement with the CRSG to enable WWF-UK to conduct the feasibility studies and draw management plans for the forest conservation project (Ebin, 1992). Later that year, the Federal Government Council of Ministers approved the Cross River National Park, which included both the Oban and Boshi-Okwangwo areas. The management plans for the Project were drawn up in 1990 after socio-economic studies were completed in the Oban and Boshi-Okwangwo areas. However, the federal law to legally establish the National Park and enable the commencement of operations did not get through the judicial process until August 1991, when Decree No. 36 of 1991 (National Parks Decree) was formally promulgated (Federal Republic of Nigeria, 1991b).

The design of both Divisions of the Cross River National Park was guided by the theoretical principles and philosophy of conservation-with-development discussed in Chapter 1. However, the conservation issues, problems and objectives differed somewhat between the two divisions (Okali, 1992; Caldecott, 1996). The Oban and Okwangwo Divisions constitute about two-thirds of the forest reserves in Cross River State or 21% of the total forest area of Cross River State (Ebin, 1992). The Oban Division is significant for conservation with respect to its forest resource and the Okwangwo Division has an added importance for species protection, particularly because of the presence of gorillas (Ite, 1995; Oates, 1999). The Okwangwo Division contains more habitat types than the Oban Division. However, this might present an unrealistic picture of the relative species richness of the two Divisions, as it seems quite likely that the forests in the Oban Division are richer than those in the Okwangwo (White, 1990). Without doubt the CRNP is the largest, most diverse protected area in West Africa. Its value in the international context is further enhanced because it is contiguous with the Korup National Park in Cameroon (see Figure 4.1).

With an initial seven-year work programme and budget (Caldecott *et al.*, 1989, 1990a), the CRNPP was designed to benefit from several sources of funding. These included a US $21 million European Community (EC) funding under the Lome IV Convention and another DM 10 million from *Kreditanstalt fur Weideraufban* (KfW) - a German Development Credit Agency

(Ashton-Jones, 1992; Ebin, 1992). In Cross River State, WWF-UK operated as the agent of the NCF, which facilitated WWF-UK in Nigeria. It provided some aspects of the technical and managerial expertise for both Divisions of the National Park and was solely responsible for the management and execution of village development projects in the Okwangwo Division of the Park (see Chapter 8). The management of the Oban Division was contracted out in 1994 by the EC to a consortium of international consultants, including *Organisation et Environment* of France and GKW Consult of Germany but led by Hunting Technical Services of the United Kingdom. However, this arrangement was suspended in 1996 when EC sanctions on Nigeria were implemented following the execution of Ken Saro-Wiwa, an environmental and human rights activist.

4.3 Rationale for the Establishment of Okwangwo Division

Specific interest in a conservation project for the area presently constituting the Okwangwo Division (Figure 4.2) begun early last century. The area was protected by the 1930 Reserve Order, and was converted into a game reserve in early 1940. In 1947, the Boshi Game Reserve was created, regulating hunting and trapping rights of indigenous communities. In 1956 the Boshi Extension Forest Reserve was proposed as a gorilla sanctuary by E. W. March of the defunct Government of the Eastern Region of Nigeria (GERN). Following Nigerian independence in 1960, the protection of the area as a game reserve continued under the auspices of the then Eastern Region and the South Eastern State Governments. In 1964, the GERN and FGN proposed the establishment of the Obudu Game Reserve comprising the Boshi, Boshi Extension and Okwangwo Forest Reserves. With the renaming of South Eastern State as Cross River State in 1976, the Okwangwo Forest Reserve was gazetted and incorporated into the list of Forest Reserves under the Eastern Nigerian Forest Service Laws and Regulations of 1981 after earlier gazettement in 1966. These series of legislation were intended to provide protection to the forest in terms of tree harvesting and all forms of land use including farming. It also granted certain rights to the inhabitants of local communities, including the rights to hunt and fish, and collect minor forest products (Shuerholz *et al.*, 1990).

After the Nigerian Civil War (1966-1970), John Oates, a biological anthropologist, advocated wildlife and habitat protection programmes in the Obudu and Oban forest areas (see Oates, 1999). Duncan Poore came into the scene in the late 1970s and advocated the Boshi-Okwangwo as a World Heritage Site (Caldecott *et al.*, 1990a). The early 1980s were characterised by continued and strong advocacy for the establishment of a National Park in the area under review, especially following the confirmation in 1983 of the presence of gorillas in the Mbe Mountains complex. The late 1980s ushered in the interest of conservation agencies both at the national and international levels. For example, the Nigerian Conservation Foundation advocated sanctuaries, integrated rural development, gorilla-based tourism and trans-frontier co-operation with Cameroon. The International Council for Bird Preservation (ICBP) made a case for a National Park status for the Boshi-Okwangwo Forest Reserve. In 1987, the area was specifically recommended for protection by the World Conservation Union (IUCN, 1987). More significantly, the Mbe Moun-

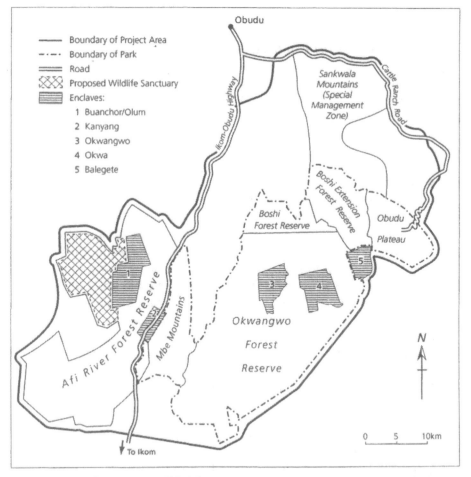

Figure 4.2   Okwangwo Division

tains complex (see Figures 4.2 and 4.3) became a focus of international conservation attention soon after the 'rediscovery' of gorillas (Harcourt *et al.*, 1989).

In the final analysis, four factors influenced and reinforced the decision to include the Okwangwo Division in the Cross River National Park project framework (Caldecott *et al.*, 1990a). The first was regional watershed protection. The Okwangwo Division in Nigeria and Takamanda Forest Reserve in Cameroon comprise one contiguous tropical rainforest system. Several of the major tributaries of the Cross River (e.g. Oyi, Okon) rise in the forest of the Division before flowing into Cameroon where the Cross River itself originates partly as drainage from Takamanda.

The second reason was the prevention of savannisation. The forests of the Okwangwo Division were vulnerable to fire due to the seasonal rainfall patterns characteristic of the area. In this respect, the protection of the Division was seen as representing an international priority in preventing the transformation of tropical moist deciduous forest to woodland savanna and degraded grassland.

The third reason hinged on the protection of biodiversity: The Division possesses a continuous progression of natural habitats between about 150m and 1,700m in elevation, the montane forest at the upper end (northern fringes) of this system having been destroyed by conversion to grassland.

The fourth reason focused on the potential for tourism. The Division had been widely reported (Harcourt *et al.*, 1989; Caldecott *et al.*, 1990a; 1990b, Oates *et al.*, 1990; White, 1990) as containing West Africa's only population of gorillas and up to a third of Africa's total number of primate species. These include two of the rarest species: the drill and Preuss' guenon (Chapter 5). The successful development of gorilla-oriented tourism in East Africa tended to suggest that the gorilla in the Okwangwo Division could be a valuable tourism asset. This is yet to be proven for the Division.

4.4   The Okwangwo Division, Cross River National Park

The Okwangwo Division of the Cross River National Park is adjacent to and contiguous with the Takamanda Forest Reserve in Cameroon (Figure 4.1). Major access to the Division (Figure 4.2) is provided via the Ikom-Obudu Highway (IOH), and the trunk road from Obudu Town to the Cattle Ranch. Several feeder roads also connect the peripheral park villages with the IOH. The Division can be reached by road within three hours from Calabar, the capital of Cross River State.

The Okwangwo Division covers approximately 1,000 km². It

Table 4.1   Forest Reserves in the Okwangwo Division

Forest Reserve	Original Order Number	Amended Order	Enclave
Okwangwo	Order No. 53 of 1930	E.R. No. 6 of 1952	Okwa, Okwangwo
Boshi	Order No. 37 of 1951	E.R. No. 4 of 1952	None
Boshi Extension	E.R. 279 of 1958	E.R. No. 294 of 1958	None

Source: Okali (1990)

comprises the former Boshi, Okwangwo and Boshi Extension Forest Reserves, including all the existing inhabited enclaves within the Reserves (see Table 4.1). For the purposes of management, the Division was divided into five management zones (Caldecott *et al.*, 1990a). These are:

- Conservation and Tourism Zone (with no hunting or gathering but supervised access for recreation and research):
- Traditional Use Zones (for supervised gathering of forest products by designated local communities):
- Recuperation Zones (areas to be allowed to recuperate following human use):
- Obudu Plateau Zone (the Obudu Plateau); and
- Special Use Zones (areas to be used in other ways, e.g. infrastructure development).

Each of these zones had different management and development objectives. The use of zones in this way seems to have been derived from the general principles of the concept of biosphere reserves (Adams, 1990).

Of particular significance to this book (Chapters 7 and 8) is the Mbe Mountains complex (Figure 4.3), a conservation and tourism zone of Okwangwo Division. It comprises approximately 11% of the total area of the Division and was been designated for total protection of the forest and wildlife as well as for controlled access by tourists. The complex is bounded in the east by the Okon River, in the south by the farmlands of the villages of Abo Ogbagante, Abo Obisu and Abo Mkpang, in the west by the Ikom-Obudu Highway and in the north by the farmlands of Wula, Bokalum and Bamba. It is a community-owned land sandwiched between the old Okwangwo and Afi River Forest Reserves. The Mbe Mountains complex is surrounded by several villages served mainly by two feeder roads. These are the road from Wula 1 to Bamba in the north and the road from Abo Ogbagante on the Ikom-Obudu highway to Bashu communities near the Cameroon border in the south (see also Figure 7.1). The feeder roads in the field area are highly seasonal

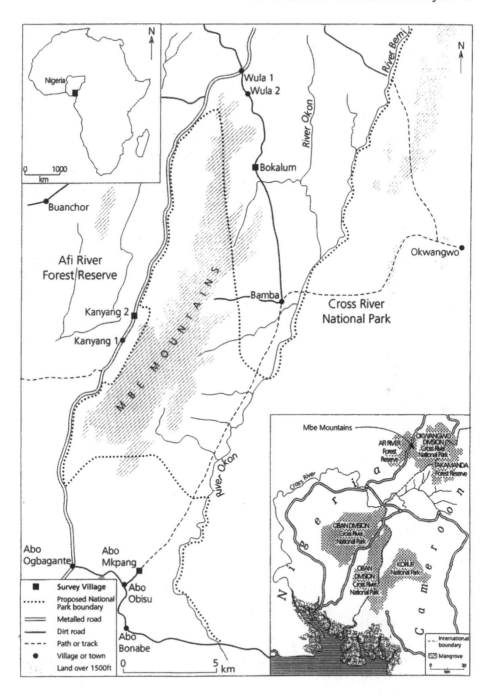

Figure 4.3　Mbe Mountains Complex

and poorly drained. They become slippery and impassable during the rains indicating the clayey or silty texture of the soils.

The Mbe Mountains is the most accessible sector of the Okwangwo Division as it borders the Ikom-Obudu Highway, the major access to the Park. The highway enhances an excellent view of the canopy forests between Abo Ogbagante and Wula 1. The Park boundary reaches the road from the east approximately 5km north of Abo Ogbagante. Where the National Park meets the road, advanced succession forest lines the boundary for about 6km to the north. The Mbe Mountains complex appears to have the most potential for nature walks and wildlife viewing. Although the central hills provide some fine views, hiking in them is quite strenuous.

It is important to note that the boundaries and extent of the Okwangwo Division were based on the boundaries of the existing Okwangwo and Boshi Extension Forest Reserves and not on the recommendations of the management plan by Caldecott *et al.* (1990a). Both the Mbe Mountains and the Obudu Plateau were omitted from the legal definition of the Okwangwo Division as reflected in the National Parks Decree of 1991 (Federal Republic of Nigeria, 1991b). These areas were left out of the decree because they were not surveyed nor gazetted as belonging to the government (Ite, 1995). Ideally, this should have been done prior to the presentation of the plan to the Federal Government for approval. The lack of time and funding for the preparation of the project documents served as the key limitation in this respect. Since 1991, significant efforts were made by the management of CRNP to resolve this issue, with varying levels of success (Barker, 1995, 1996, 1997 and 1998; Caldecott, 1996).

4.5   Management and Development of Okwangwo Division

From the outset, threats to the integrity of the Division were considered by the park planners to derive mainly from hunting and farming by the local population around the Okwangwo Division. However, in many respects these fears have remained largely unjustified (Ite, 1996, 1997; Caldecott, 1996; Ite and Adams, 1998). The communities adjacent to the National Park depend upon access to the resources of the Division for their livelihoods (Chapter 5). For example, the two enclave communities of Okwa and Okwangwo and those surrounding the Mbe Mountains were identified by WWF-UK as the most critical sources of threats to the physical integrity of the forest and consequently, the conservation area. This stemmed from perceptions of their high degree of dependence on and use of resources in the area designated as the National Park.

Based on the understanding that the Park area and its surrounding communities are interdependent and therefore must be managed coherently, three complementary mechanisms were designed to protect and develop the Park area. These are, first, a park management service responsible for supervision, law-enforcement and public relations. Second, a development programme aimed at linking conservation of the protected forest areas with local socio-economic development by assisting the surrounding villages to make more profitable and sustainable use of their land. Third, a series of economic mechanisms designed to enable the local (support zone), state and national economies to benefit from the commercial development of its genetic resources (Caldecott *et al.*, 1990a).

The three mechanisms described above possess the characteristics of 'conservation-with-development' discussed in Chapter 1, and fit the general framework as regards their conception, design and application (Stocking and Perkin, 1992). Such strategies represent significant effort to replace the old confrontational approach between park managers and local people with one that attempts to address the needs of local communities. As noted in Chapter 1, several studies (Hannah, 1992; Wells *et al.*, 1992; Barrett and Arcese, 1995; Ite and Adams, 2000) have shown their ineffectiveness in addressing conservation and development issues simultaneously.

4.6   Forest Conservation in Okwangwo Division: Aims and Intentions

A key planning principle during the process of establishing the Okwangwo Division had been that conservation and development are, ultimately, complementary activities (Caldecott *et al.*, 1990a). The Okwangwo Division was expected to provide an opportunity to achieve both the conservation goal of maintaining a globally important ecological area intact, as well as the development goal of improving the economic and social welfare of the people to be affected by the park project. This involved the application of a 'carrot' of public benefits and a 'stick' of regulations and enforcement in the villages within and bordering the Park. The mechanism for the realisation of this goal was described by Caldecott *et al.* (1990a) as the support zone development programme (SZDP).

Based on the concept of the SZDP, the target villages, their farmlands and communal forests were defined as 'support zones'. Within the support zone a combined package of incentives and disincentives were to be applied to encourage local people to participate actively in the protection and development of the National Park, and to change their land use practices to a more sustainable, agroforestry-based systems. The SZDP placed considerable emphasis on an effective agricultural extension service as the key developmental operation in the support zone of the National Park and was closely linked to an education and information programme. Other components of the SZDP included wildlife management, rural road maintenance, and the development of small-scale industries. The objectives of the extension work included improving management practices on the farms; introducing better varieties of existing crops, and maximising production on existing agricultural areas. Others were to reduce the requirement for new agricultural land, increase the productivity of local domestic livestock, in particular the pigs and goats, as well as develop the intensive culture of fish in fishponds (WWF-UK, 1992). To benefit from the package of integrated rural development, communities were required to register with the SZDP and to pledge to respect the boundaries of the national park (see also Chapter 8).

According to the management plan for the Okwangwo Division drafted in 1990 (Caldecott *et al.*, 1990a), the incentives of the SZDP would include directed credit, advice, grants and planting materials. All incentives were to be supplied on condition that the beneficiaries would respect the National Park boundaries. The incentive structures were designed to improve the financial environment of the support zone. For example, provision was made for a revolving credit fund to relieve the support zone farmers from the inhibiting effects of credit rationing, provided they invested in appropriate projects determined and approved by the park management (Chapter 8). Farmers in the area had poor access to credit and this was seen as the key restraint on their ability to improve or change their systems of land use. Other financial incentives envisaged by the SZDP included a village conservation and development fund as well as a crop loss compensation fund.

On the other hand, the disincentives for encroachment on the forest varied. These included the reduction of regular allocation grants, suspension of SZDP registration, refusal of loans to individuals and withdrawal of specific privileges to exploit non-timber forest products. In accordance with conservation objectives for the Division, the priority with which development assistance would be provided to support zone villages was based on the proximity to, and the perceived impact of each village on, the area constituting the National Park (Ite, 1995). The primary intention of the SZDP was to create a 'buffer zone' around the protected area in a bid to cushion the apparent threats (real or imaginary) to the Park. The view was to keep major human impacts at a greater distance than a conventional physical boundary would do. The SZDP (as planned) therefore placed emphasis on the mutual dependence between the Park and nearby communities through a rural development programme (Chapter 8), rather than a defensive posture of the park against the people (Sayer, 1991b). However, the definition of the support zone in terms of villages rather than a discrete area implied that it could not be shown on a map, with significant implications for the accurate determination of the specific area of its influence (Ite, 1998).

It is important to note that the management plan for the Division was completed in 1990 when there was no legal instrument defining the National Park. Between the feasibility study and the wider activities to be funded by the European Community as envisaged by the management plan, an Intermediate Phase Project (IPP) was initiated in the Oban and Okwangwo Divisions of the park project. The objective was to keep the ICDP concept alive, pending the promulgation of the National Parks decree and the full take-off of the park project. The IPP was spearheaded by WWF-UK and facilitated by the NCF, with funding from various sources, including the WWF-UK, the UK Overseas Development Administration (now Department for International Development), the Cross River State Government, the Federal Government of Nigeria, the Shell Petroleum Development Company of Nigeria and the Ford Foundation. Overall, the IPP had the goal of establishing wide local community support and developing a project institution. The implementation objectives of the IPP included village liaison, environmental education, park protection and village and institutional development (Ashton-Jones, 1992; Marshall, 1993a; 1994a; 1994b). The extent to which the SZDP achieved its overall objectives is evaluated in Chapter 8 of this book.

4.7 Conclusion

The Cross River National Park project is a direct manifestation and transformation of global thinking on TMF conservation into local practice and action. It is evident that attempts at the conservation of the forests of the Okwangwo Division of the CRNP dates back to the 1930s. It took nearly half a century before the proposals to constitute the Okwangwo forests into a National Park became a reality. There is no doubt that international conservation interest groups orchestrated the establishment of the Okwangwo Division. On the other hand, it could be argued that the park project was designed in response to the conflicting, divergent needs and demands of conservationists, international development agencies as well as those of the Federal and Cross River State governments and the local communities.

Part III of this book examines the local realities of tropical forest loss and conservation in Okwanwgo Division of the Cross River National Park project.

PART III:
LOCAL ACTION: FOREST LOSS AND CONSERVATION IN OKWANGWO DIVISION

5 Environment and Development Themes

5.1 Introduction

This chapter examines aspects of the physical environment, social and economic development in the Okwangwo Division of the Cross River National Park. It describes the salient physical attributes of the Division with particular reference to the topography and geomorphology, geology, hydrology and drainage as well as soils. The chapter also reviews the ecology of the Division, especially the natural vegetation and wildlife resources. From the development perspective, the ethnography, population characteristics, local social and political administration, land tenure system and traditional land use and economy of the Division are examined.

5.2 The Physical Environment

Cross River State has one of the most diverse landscapes in Nigeria. This is also reflected in the topography of the Okwangwo Division, which is extremely varied with relatively flat area, steep hills, escarpments and moderately steep river valleys. In general, the land within the Okwangwo Division rises northwards and eastwards from the surrounding lowlands. The elevation range from about 45m at the Okon River near Bashu to over 1,900m at the summit of the Sankwala/Obudu Mountains, west of the Cattle Ranch. In general, the Obudu Plateau, Mbe and Sankwala Mountains represent areas with marked relief, which could attain an altitude of over 1,700m above sea level, while lowland undulating plains, interrupted by small hills and ranges, characterise the Okwangwo Forest Reserve. Geologically, acid crystalline metamorphic and granitic rocks dominate the Okwangwo Division. These rocks are quartz-feldspar-rich with varying amounts of biotite, hornblende and garnet (Shuerholz et al., 1990; Petters, 1993). Groundwater depends exclusively on discontinuities in the rocks. A deep water table is associated with joints; fractures or shattered zones but the aquifers are small and localised. Consequently, the potential for groundwater in the field area is low. Therefore, runoff constitutes the major source of water supply in the Division.

Hydrologically, Okwangwo Division is well served and dissected by

large rivers. The Okon River, a tributary to the Cross River, is the most prominent in the Division. Others are the Bemi (or Emi) draining into the Okon,
Anyuko and Mache feeding the Oyi, Aseche which flows into the Mbe river
originating from the Mbe Mountains and joining the Okon. These rivers are
perennial, contrary to the Kasagache, Echie, Njujjua, Matche, Katche, Anyuko,
which are ephemeral. The complex topography cause sharp changes in direction of river courses in the higher areas (Caldecott *et al.*, 1990a). Using the
FAO systems of soil classification, Schuerholz *et al.* (1990) described the
soils in the Okwangwo Division as Ferrisols, Acrisols and Drystic Landisols.
The soils portray the distribution of two classes (lithosols and lateritic soils),
differentiated on the basis of morphology and profile development. The
lithosols are shallow, stony soils occurring on steep slopes where profile development is retarded due to erosion. They are composed largely of fragmen-
tary rock materials. These soils are sandy, infertile, increasingly stony, shallow and erodible on steeper slopes (Ite, 1995).

5.3   Vegetation and Biodiversity

The character of vegetation in the Okwangwo Division changes from place to
place. This reflects the interaction amongst human activity, rainfall seasonality
and total, elevation, aspect and substrate (Caldecott *et al.*, 1990a). Vegetation
in the Division consists of rainforest and northern guinea savanna, which is
transitional in nature. As such, the whole of the Division can be regarded as
lying within the West African forest-savanna transition zone. Although the
three Forest Reserves in the Division (i.e. Okwangwo, Boshi and Boshi Extension) lie wholly within the forest zone, the northern limits of the Park are
within the forest-savanna ecotone. This is a tension zone where the change
between forest and savanna depends delicately on the balance between the
physical factors of soil, climate, and perhaps elevation on the one hand, and
the human factor of land use on the other.

Forest vegetation in the Okwangwo Division is a mixture of gallery or
transition forest, primary forest, secondary high forest and outright savanna
woodland. The change from forest to savanna woodland is sometimes made
abrupt by changes in farming practices (Chapter 6). As Okali (1990) noted,
while boundaries between a patch of forest and an adjacent savanna woodland may be readily defined, the transition zone is really a mosaic of patches
of forest and savanna, with patches of each vegetation type lying well within
zones of the contrasting vegetation. Usually forest outliers occupy wetter areas such as valleys within savanna woodland, while shallow soils or particularly intensive or prolonged land use leads to savanna patches within forests.

The issue of whether savanna is advancing on forest, or vice versa, in a given area, can be only resolved by a long-term study (see, for example, Fairhead and Leach, 1994). This is outside the scope of this book.

In the Mbe Mountains complex (Figure 4.3), the vegetation is an example of Guinea-Congolian high forest growing in a strongly seasonal climate. Climax forest covers at least 150 km^2 of the mostly mountainous terrain with maximum elevations of 900 m above sea level. In the southern part of the Division, a large proportion of the biggest trees is *Brachystegia nigerica*, with huge spreading crowns reaching a height of 45 metres. According to Oates *et al.* (1990), this specie is known to exist only in Nigeria. Other common trees in the area are "ironwood" (*Lophira alata*) and *Entandrophragma utile*. In most areas the forest canopy is closed and little disturbed by agricultural and logging activities. Secondary forests are common around the various villages of the study area, obviously due to biotic interference e.g. clearing, burning, illegal farming, and timber exploitation.

The diversity of the fauna of Cross River State reflects the great species richness of the vegetation, although there is an increasing paucity of species as rainfall decreases to the north and west. In the Okwangwo Division, the primate fauna is diverse. Based on direct evidence (sighting or hearing animals, observing nests, or seeing animals shot locally by hunters), Oates *et al.* (1990) confirmed the presence of at least 11 species in the forests of the Okwangwo Division (Table 5.1).

According to the World Conservation Union (IUCN), the drill, red-eared guenon, and Preuss's guenon are endangered species. The Okwangwo

Table 5.1   Some Primate Species of the Okwangwo Division

English Name	Scientific Name
Gorilla	*Gorilla gorilla*
Chimpanzee	*Pan troglodytes*
Drill	*Mandrillus leucophaeus*
Putty-nosed guenon	*Cercopithecus nicitans*
Mona guenon	*Cercopithecus mona*
Red-eared guenon	*Cercopithecus erythrotis*
Preuss' guenon	*Cercopithecus preussi*
Needle-clawed galago	*Euoticus elegantulus*
Allen's galago	*Galago alleni*
Demidoff's galago	*Galagoides demidoff*
Potto	*Perodicticus potto*

Source: Oates *et al.* (1990)

Division probably contains 17 primate species, or nearly a third of Africa's total. It is important to note that although the primate fauna is diverse, the animals themselves are far from abundant in the area. During a survey of 471km of potential primate habitat at slow walking speed, Oates *et al.* (1990) had only one clear view of a monkey. This is in stark contrast with 108 visual encounters in a 373km survey on the island of Bioko in Equatorial Guinea, where a similar but not identical primate community is known to occur. The Okwangwo Division has been known to contain the only population of gorillas in West Africa. Oates *et al.* (1990) indicated the presence of an estimated population of 110 gorillas, with 150 as the more realistic upper limit of gorilla population in the area. The largest sub-populations were found in the Mbe Mountains complex and Afi River Forest Reserve.

Many mammal species occur in the Okwangwo area in addition to the primates. White (1990) recorded over 30 species of mammals (excluding rats, mice, shrews or bats). Again, direct sightings of mammals are infrequent in many cases probably due to hunting pressures. Significant bird populations are also found in the Okwangwo area, including red-headed (or grey-necked) rock fowl (*Picathartes oreas*), green ibis (*Bostrychia olivacea*), violet-backed flycatcher (*Hyliota violacea*) and the black guinea fowl (*Agelastes niger*). Some of these birds, found nowhere else in Nigeria, are threatened.

5.4   Ethnography and Population

Except for the Sankwala Mountains, the Boki people inhabit most of the Division. One account of Boki origin suggests that they lived in Cameroon as a homogeneous group in one contiguous geographical area, speaking the same language or mutually intelligible dialects. Ewah (1986) cited by Shuerholz *et al.* (1990) referred to the Boki people as semi-Bantu stock, who expanded into new lands and rivers for farming, hunting and fishing. Haig (1949), cited by Shuerholz *et al.*, (1990) traced the origin of the Boki to one ancestral founder who established Danyere, and whose family later grew into Abu, Osokom, Boje, Irruan and Eastern Boki clans.

Regardless of the debates on the origin of the Boki people, it is important to note that Cross River State comprises a fairly large, yet undetermined number of ethnic groups (Abasiattai, 1987). However, very little scholarly research has been done or information published for general knowledge about many of these ethnic groups, especially those north of the Old Calabar Province. This area is made up of various ethnic groups usually given the generic name *Ekoi,* although collectively they are less than a million people. They include the Ejagham, Bekwarra, Bette, Bokyi, Agwagwune, Bahumono and

Yakurr. This diversity in the Upper Cross River languages has resulted in their being grouped into as many as 25 languages clusters (Abasiattai, 1987; Essien, 1987). However, the three major ethnic groups in the State represented by the three principal linguistic groups are Efik, Ejagham and Bekwarra. The people in the Mbe Mountains complex are Boki-speakers, belonging to the Cross River: Bendi language group and Eastern Boki group of villages. Boki villages closest to the Division's forest boundaries are two Bashu villages, three Abo villages, two Kanyang villages, Bamba, Bokalum, Wula, Butatong, Okwangwo, Kakwe Bebo and the 12 villages of Bumaji. In contrast, people on the western side of Afi River Reserve belong to the Western Boki group of villages.

The 1991 population census of Nigeria suggested that approximately 277,441 (15%) of the population of Cross River State reside in Boki, Obanlikwu and Obudu local government areas (LGAs), where the Okwangwo Division is located. The 1990 'Masterplan' for the Okwangwo Division estimated the population of the 66 'support zone' villages at approximately 36,000 people (Caldecott *et al.*, 1990a). This varied from 2,600 people at Bateriko to 35 people in Jato on the Obudu Plateau. The mean village size was put at 540 individuals with the median at 480 people (Caldecott *et al.*, 1990a). A revised work programme and funding proposal for the Okwangwo Division (WWF-UK, 1992) specifically focused on 35 out of the 66 villages and gave a population figure of approximately 16,200 for the selected villages. Ite (1995) estimated that at least 6% of the population of the three LGAs (i.e. Boki, Obanlikwu and Obudu) would be directly affected by the development and management activities of the Okwangwo Division. The lack of precise land use data precluded the calculation and derivation of the population densities for each LGA. This could have provided a basis for extensive quantitative analysis of the implications of population pressures on natural resources, especially the forest of the Okwangwo Division.

5.5 Village Administration: Political and Social Aspects

There are variations in the traditional instruments of power and the modes of political administration in Cross River State, which vary from one group to another. However, sociological and anthropological studies have revealed common features in the traditional political system of the peoples of the State (Ekong, 1987). These include:

- Absence of a single centralised government with a paramount ruler such as were found among the old empires in northern and western

Nigeria:

- Organisation of the society on the basis of the segmented unilineal principles:
- Strict adherence to primogeniture and sometimes leaving social power in the hands of elders:
- Perception of social power as ultimately inherent in or derivable from supernatural sources and the use of secret cults as the custodian of such powers and:
- Administration by a portfolio system, the use of local formal organisations in routine local administration and decision making by consensus.

In Cross River State, socio-political powers are traditionally exercised in distinct hierarchical stages, with the immediate (extended) family unit as the base. The village is the most significant unit for the organisation of economic, social and religious life. Boki villages are headed by a Chief elected by the council of elders (*Bachi-Ikpe* in Boki), and confirmed by the Local Government authority. Tribal traditions are protected by several societies each with a number of taboos. There is no doubt that the cultures of the peoples of Cross River State are well defined, being strongly governed by the rule of law, especially community laws, norms and regulations (CRFD, 1994). Community laws governed inheritance, land tenure, government hierarchy, lines of authority and respect, marriage and kinship, property rights and spiritual laws. These significantly influence the general behaviour of both community and individuals. However, due to former colonial and modern pressures, local culture, traditions and social mores have undergone significant changes. Long established social structures and lines of authority are in the process of breaking down and some are clearly no longer relevant.

There are other instruments of administration in the villages. These include formal social groups such as Age Grades, Youth and Women's Associations as well as secret cults. Village development committees include those responsible for education, health, markets, village land use and allocation. One or more sub-committees regulate the exploitation of natural resources in the villages. Indigenes have the full right to forest use based on well-established principles. Non-indigenes intending to exploit community forest for timber and other forest products have to negotiate the terms and conditions with the Chief as well as the appropriate village sub-committee(s).

Apart from the regulation functions, these committees are responsible for generating revenue for the village from the exploitation of resources from the community forests. For example, fees are collected from the harvesting, purchase and evacuation of non-timber forest products such as bush mango and salad, from the villages. Timber dealers could directly buy from the villages the right to log specified species in the community forest. Villages are entitled to royalties (about 30% of the stumpage fee paid by the logger) from the Cross River State Forestry Department if logging takes place in Forest Reserves on community land. The revenue accruing from the activities of the relevant resource management committees are utilised in village development projects, for example, education, markets, road construction. This demonstrates the significance of community-level development and resource management organisations in developing countries (see Okoye, 1988; Ite and Adams, 2000).

5.6 Land Tenure System

Land tenure in Nigeria involves sets of rules and regulations for managing natural resources and the environment. Tenure systems are dynamic and respond to socio-economic and political changes put in place for resource utilisation (Osemeobo, 1991). They are, however, not mono-specific and vary from one rural community to another (Udo, 1990), but are essentially pivoted by two broad systems of communal and individual ownership of land. Land tenure system influences the use of land for economic and social developments. Yet land use determines the extent to which a resource could be conserved and the level of conservation attainable (Osemeobo, 1991).

The traditional land tenure system in southern Nigeria (including Cross River State) is based upon several concepts, most of which are common to tropical Africa. These include the following:

- Land is the joint property of the community and there is no basic concept of individual right of permanent ownership:
- The right of an individual to obtain land for farming is contingent on membership of a community either by birth, adoption or marriage:
- There is a single traditional custodian of the group's rights. It is the responsibility of the custodian to exercise control over the land and the group and allocate the former in accordance with traditional laws and customs:
- Continuous right to land is contingent on its active use.

However, the interpretation and implementation of the above principles varies among ethnic groups. Each community in the Okwangwo Division lays claims to the forest area surrounding the village. The protected forests, the main source of land for farming are mostly communally owned.

Members of the community wishing to utilise fresh (primary) forest for farming can clear as much land as they desire. In communities with inadequate supply of forest areas, the community land is usually shared out to families/households. The ownership of the land by individuals is normally derived from initial clearances of portions of virgin community forest. The descendants of such individuals can inherit the land. Thus in all cases, land owned by an individual or family remain so owned and can be inherited (Caldecott *et al.*, 1990a). There is therefore a mixture of communal and individual ownership of land in the communities of Okwangwo Division. Data collected by Ite (1995) showed that two-thirds (63%) of the farm plots owned by households were obtained through initial clearance of portions of the community forests by the respective households. This tallies with the findings that 62% of the plots were obtained by the households free from any form of charge. However, although the data also showed that households, through inheritance, obtained one-third of the farm plots in the villages, it is important note that no individual farmer has legal title to the land that he or she occupies for agricultural purposes. Nonetheless, acquisition of land through forest clearance does give the farmer right to continual access to the plot. No other person can cultivate this land without the permission of the person who first cleared it. The same principle applies to land inherited from the extended family. No other person has the right to occupancy even if the land is cleared, cultivated for one season and/or left fallow for an extended period of time.

Land tenure arrangements are different for non-indigenes of the villages in the Okwangwo Division. Immigrants to the area are required to rent land from the community or individuals for temporary use and are not entitled to permanent ownership. Before land is leased to immigrants, they are required to formally register with the host community in cash and/or in kind (e.g. providing drinks). On being allocated farm plots, yearly dues have to be paid to the village council or the individual (depending on who leased the plot) as long as the immigrant remains in the village. In most cases, these dues are paid even if the plots are under fallow. If migrants marry indigenes of the village within the period of their residence, they could be given land free by the in-laws, although this does not necessarily provide an automatic right to other forest exploitation activities, e.g. hunting.

5.7 Traditional Resource Use and Economy

Agriculture and forestry constitute the mainstay of the Cross River State economy, employing over 80% of the state's labour force and contributing about 70% the GDP of Cross River State. Small-scale (household level) farming is the main occupation of rural people in the State. It is the main route to socio-economic development for rural people. The State is also a major exporter of timber and non-timber forest products (NTFPs) to other states in Nigeria. In the Okwangwo Division, the land use and the economy of the are largely agrarian, but in general terms, small-scale farming, hunting, trapping and collection of forest products are of greater significance.

There is no doubt that the procurement of bushmeat for consumption and sale is an important traditional economic activity in Nigeria (see for example, Adeola, 1992). In the Okwangwo Division, there is comprehensive participation in hunting by all the communities (Ite, 1995). Animals in the forest are both hunted and trapped, while in the farms they are trapped to protect crops. Hunting is a year-round activity but peaks in the rainy season when it is easier to move quietly on the forest floor. The hunting radius from the various villages could range from 5 to 20km and excursions lasting more than a day are quite common (Oates *et al.*, 1990). The animals mostly shot in the high forest are porcupine, duiker, guenon monkeys and bush pig. In other areas, the most prominent and common prey species are, cutting grass, rock hyrax and bushbuck. Bushmeat is highly regarded and local production does not satisfy demand, resulting in high prices being paid where a market can be reached. Only 50% of the bushmeat caught in the Okwangwo Division are sold outside the area. Fishing is relegated to a secondary, occasional activity. Livestock rearing, especially goats, sheep and cattle (*muturu*) are practised to a limited extent.

Non-timber forest products (NTFPs) help to stabilise income and employment in the various farming communities of Cross River State (Omoluabi, 1994; Omoluabi and Abang, 1994). The benefits of trade resulting from the NTFPs of Cross River State origin have been significant. Within the context of the Okwangwo Division, the wood products and species gathered by local communities from the forest include poles (e.g. *Harungana madagascariensis*) and fuelwood. Others are chewing stick (with the *Massularia acuminata*, *Garcinia manii* and *Carpolobea lutea* as the most important sources), tool handles (e.g. *Baphia nitida*) and stem sponge (e.g. *Momordica angustisepala*). Bundles of sticks from several genera are collected and exported to northern Nigeria to be used by cattlemen for herding and building huts. The most significant NTFP in the Okwangwo Division are the two varieties of bush mangos (*Irvingia spp.*). Rattans (cane palms) are also collected in many communities on a small scale. Rattans are used as tying material in house construction, as well as for the fabrication of household goods such as *garri* sieves, baskets and furniture. Households in the Division derive considerable cash incomes from the sale of NTFPs. Although estimates by households vary widely and sometimes can be intentionally inflated (Ite, 1995), they provide an indication of the dependence of local people on these products. The highest income generating activity is the collection of bush mango, followed by harvesting of leafy vegetables otherwise known as 'salad' in the study area.

There are several off-farm income sources open to households in the Division. Many households are engaged in trade and commerce, with the most common commodities being agricultural produce (e.g. banana), non-timber forest products (e.g. bush mango and salad) and bushmeat from hunting. The marketing of these commodities are confined to the villages principally due to the poor state of transport development in the area. The prices of these commodities are determined to a large extent by the buyers and their agents who have to travel to the villages to negotiate the prices and evacuate the products for sale in the urban centres.

Less than one-third of households derive income from timber extraction from the forest, although two-thirds of all households are involved in some form of timber extraction from the community forest. This suggests that approximately 50% of the households extract timber for domestic as opposed to commercial purposes. Furthermore, it reinforces the observation in Chapter 3 that large-scale timber extraction in Cross River State had been hampered by lack of access routes for evacuation. Survey data (Ite, 1995) also show that several economic activities are insignificant sources of off-farm household incomes. These include wage labour, livestock raising, labour migration, artisanal production, civil service and remittances.

5.8   Development Infrastructure

Many indicators of socio-economic development are either totally absent or grossly inadequate in the villages of the Okwangwo Division. For example, the villages depend on local streams for water supply. During most dry seasons, some of the villages (especially Abo Mkpang) experience acute water supplies. This motivated the rationing and allocation of water and the regulation on the times which water could be fetched from the local streams. Water-related illnesses and deaths have been reported as an annual incident in the villages especially in Abo Mkpang (Ite, 1995), resulting in villages having to intensify local efforts in local watershed management and conservation. For example, forest clearing and farming is no longer permitted in the headwaters and banks of the community streams.

Overall, in the Okwangwo Division, households perceived poverty as the lack of disposable income and the inadequate provision of community socio-economic development infrastructure. Community development priorities include water supply, health care and nutrition (MacDonald, 1994); electricity, roads, education, and agricultural development with the most important being the grading of existing rural feeder roads. However, the implications of improved road systems for the National Park and the local economy has been a subject of great interest and concern for the conservation managers in the Okwangwo Division (Ite, 1997). In conservation circles, the fear has been that this might promote easy access to the protected forest area and intensify migration with further opening up of the forest as the consequence (see Chapter 7).

5.9   Conclusion

From an environmental perspective, the vegetation of the Okwangwo Division is a mixture of gallery or transition forest, primary forest, secondary high forest and outright savanna in some places. There is a diversity of wildlife, with the gorilla as the most important for the purposes of conservation. From a development perspective, nearly all the households in the villages are of Boki origin, and most have been in residence in the villages for a period spanning over fifty years. The extensive forest resources and very little evidence of physical development infrastructure influenced the rate of migration into the villages of the study area. The local political system provides for various village-level organisations and institutions, which play significant roles in the management of the community resources and general socio-economic development.

6 Agricultural Land Use: Practices and Challenges

6.1 Introduction

Agriculture is a significant form of land use with environmental impact on forests, including losses of biodiversity and habitat principally due to forest fragmentation. As Wood (1993) noted, in spite of the apparent economic advantages of expanding agriculture into forest areas, the policy of converting forests in developing countries to farmlands has been repeatedly questioned.

This chapter describes the nature, patterns, practices and challenges of agricultural land use in the Okwangwo Division, Cross River National Park. Particular reference is made to the villages of the Mbe Mountains complex, where the park planners identified agriculture as a significant threat for the gorilla habitat (Chapter 4). The materials presented in this chapter provide the basis for articulating the role of farmers in forest loss (Chapter 7), and forest conservation (Chapter 8) in the Okwangwo Division.

6.2 Household Resource Base for Farming

There are differences in the distribution of farm plots among households and villages of the Okwangwo Division. These could be explained by several possible factors, including the extent of community forest available for farming; labour availability; the importance of agriculture in the economy of households relative to other forms of forest land use (hunting, gathering) in the villages, and household sizes (Ite, 1995). There is a gender bias in the access to and ownership of land. Men (husbands) own almost all the farm plots through initial clearing of the forest or by inheritance from parents (Chapter 5). Women obtain farm plots from their husbands, his relations, or their own fathers, brothers or other male relations. Although husbands claim ownership of the plots, their wives or other household members end up cultivating them. Thus the ownership and subsequent cultivation of farm plots are not entirely vested in one individual member of the households. This reflects to a large extent the prevailing land tenure system in the area which makes provision for the temporary and permanent transfer of ownership of plots both within and outside the household system.

Farm sizes range from 0.2 hectares to 3 hectares depending on the crops being cultivated. Measurements in a sample of farms (Ite, 1995) suggested that, for example, the average cassava, yam and maize plot was two hectares while the average for banana and cocoyam plots was 2.5 hectares. The average size of cassava farms as a sole crop was 1.7 hectares (see also Singh, 1986; CRADP, 1986). Farm sizes vary between the villages of Okwangwo Division. The variations are influenced by several factors including the topography, labour and availability of sufficient land in the villages. Although in general farm sizes tend to decrease principally due to population growth and lack of labour, the sizes of banana farms have increased tremendously in recent times (Ite, 1997). The boom in banana production and marketing served as the prime mover for households to clear more forest to establish new farms or expand old ones. From the households' perspective, Ite (1995) found that farm sizes influences households' income. Households are apt to correlate large banana farms with increased household income, although reliable statistics are not available to substantiate these claims.

The labour force for agricultural land use in the Okwangwo Division is principally derived from three sources. These are household or family, communal or group and hired labour. The number and characteristics of household members determine to a large extent the capacity of household and family labour available for agricultural purposes. On average, there are four members (adults and children) per households available for farm work. On the basis of the occupational characteristics of household members (Ite, 1995), only 50% of the total labour potential of households are engaged in full-time agricultural production. The remaining are assumed to be part-time farmers. Overall, farming is the major occupation for adult members (18 years and above) of the households in the villages, while the majority of the younger population are in full-time education. Nonetheless, children provide much-needed assistance in the farms during school holidays.

The extensive and seasonal characteristics of the prevalent farming system in the Okwangwo Division often result in the need for extra farm hands. Households tend to utilise communal, group labour and hired labour for felling, planting, and weeding and occasionally for the harvesting of crops. It is also a common practice for households to hire labour for the evacuation of farm produce (especially bananas) from the distant farms. The division of labour by gender with respect to farm management is fairly rigid, with men and women jointly cultivating the same plot. Male tasks include felling and clearing the forest, managing tree-crops, planting cocoyam, cassava, banana and plantain. Women are more involved in digging, weeding and planting of several crops including vegetables, maize and melon.

Household capital equipment for farming in the area consists of simple hand operated tools such as hoes, axes and machetes. Limited use is made of mechanical power (e.g. power chain saws) in farming operations. Households acquire different kinds of hoes for different purposes (e.g. weeding and digging), while different knives and cutlasses are used for a variety of tasks (e.g. bush clearing, felling small trees, tapping). After harvesting, crops and fruits are transported by head-loading using baskets and basins of various sizes from the farms to the villages. In sum, agricultural production and associated processing in the area still depend to a large extent on the use of simple equipment and physical human power.

6.3   Cropping and Farm Systems

The cropping systems in the Okwangwo Division consist of mixed, multiple, and intercropping of various plants. In general farming practice are primarily constrained by terrain features and the range of feasible options limited especially on steep land. The above cropping systems exploit, as much as possible, the micro and macro characteristics of the landscape. As Anthonio (1990) observed, they sufficiently represent the culmination of decades of farming in communion with little or no extension assistance, low level technology under the constraints of a very harsh physical environment, especially in relation to the relief and topography.

Four principal farm systems can be observed in the villages of the Okwangwo Division. These are:

- Homestead or compound farms around the houses with trees:
- Cassava, yam and maize farms in secondary bush or farm fallow areas:
- Banana and plantain farms in primary or mature secondary forest areas, and
- Perennial tree-crop farms.

Homestead or Compound Farms
Homestead or compound farming is a type of land use which is of general importance in the humid tropics (Lagemann, 1977). It is characterised by annual farming with mixed sequential cropping or monoculture and scattered trees. The cultivation techniques are adapted to the needs of each plant and the microenvironment of the surface conditions in Okwangwo Division. These farms are located round the villages with several crops including cocoyam (*Xanthosoma mafaffa*), yams (*Discorea* spp.), maize (*Zea mays*) and vegetables e.g. fluted pumpkin (*Telfaria occidentalis*), bitter leaf (*Veronica amygdalina*). Many fruit trees including mango (*Magnifera indica*), guava

(*Psidium guajava*), coconut (*Cocos nucifera*), avocado pear (*Persea americana*), orange (*Citrus* spp.), pawpaw (*Carica papaya*), and Indian almond (*Terminalia catappa*), can also be found in these farms. These trees are normally planted to provide shade, fruits, nuts, edible leaves and medicinal products both for sale and household consumption. As Ite (2000a) observed, at the household level, livelihood strategies constitute the main determinant of households' decision to integrate trees in their farms. However, there is no doubt that induced innovation has a wider and significant role at the community level than at the household level in encouraging the integration of trees in homestead farms. Overall, households' adaptation to the dynamic changes in economic, environmental and social conditions constitute the dominant factors leading to the emergence of tree integration in homestead farms.

From a physical landscape perspective, a multi-storey physiognomy with perennial and annual crops is typical of the compounds, with the height of crops ranging from 0.3-25metres (see also Ruthenberg, 1976; Lagemann, 1977). The leaf canopy becomes denser, the closer it is to the ground. This reduces erosion by absorbing rainfall impact, shades the land thus reducing soil temperature, and provides for nutrient recycling through leaf litter. The leaf canopy also maintains reasonable levels of organic matter and conserves soil moisture during dry periods. Intercropping and sequential cropping (phased planting) within this farm system provides a mechanism for regulating the supply of food throughout the year and spreads the labour input over the larger planting period.

Cassava Farms
These are generally found in areas with the period of fallow not exceeding five years. Cassava (*Manihot esculenta*) is normally planted on farmland that had been under fallow after the cultivation of banana and plantain. Unlike yam, cassava cultivation is less demanding of soil nutrients and labour. As such, primary forest is rarely cleared for the cultivation of this crop. The increasing demand for *garri* (cassava flour) increases the relative price and value of this crop, thus stimulating cultivation and production. Improved varieties of cassava (e.g. HYV-77) are available to some extent in the villages, but households tend to cultivate several traditional varieties (e.g. "Panya" and "Don't Worry"). These varieties have different maturity periods, ranging from 6 to 18 months. Although improved varieties may be popular with some households due to their high yielding nature and resistance to mealybug and leaf mosaic, Ite (1995) found that there are significant problems of availability of the improved varieties as and when required by households.

For example, compared with the traditional varieties the improved varieties of cassava cannot be stored in the soil for an extended period of time after maturity. They need to be harvested after maturity and this may require the hiring of additional farm labour, with implications for household income. Furthermore, as opposed to traditional varieties, the improved varieties are not very suitable for the preparation of 'akpu', a local food item produced from fermented cassava. Even if supplied with large quantities of the improved varieties, households are more likely to cultivate the local varieties due to their known desirable qualities (Ite, 1995; Environment and Development Group, 1998). The intercropping of cassava with maize and vegetables is also a common practice in the villages of the Okwangwo Division. These are planted during the onset of the rains after clearance of fallow. They often reach maturity before cassava. Yams are also occasionally intercropped with cassava. The cultivation involves the making of mounds in areas where households considered the soil fertility level to be very high.

Banana and Plantain Farms

Bananas (*Musa sapientum*) and plantains (*Musa paradisiaca*) are the major commercial crops of the Okwangwo Division, especially in villages of the Mbe Mountains complex (Figure 4.3). Banana and plantain cultivation banana takes place at the onset of the rains in April or May. They are usually undertaken in newly cleared primary or mature secondary forest. This has been one of the factors influencing the extent and rates of forest loss in the Okwangwo Division (Chapter 7). Although improved varieties of both crops are also available; almost all households cultivate traditional varieties such as "Congo" and "Kpobata". The suckers are either collected from old farms or bought from other farmers. Most households follow a specific pattern of spatial arrangement of banana stands. The result is a variation in plant populations from farm to farm and from village to village. This depends to a large extent on the level of intercropping practised by the farmer and the effectiveness of the forest clearing exercise (Holland *et al.*, 1989).

During the first year of cultivation, banana and plantain can be intercropped with a number of annual crops such as fluted pumpkin (*Telfairia occidentalis*) and cocoyam (*Xanthosoma mafaffa*). After these crops are harvested, plantain and banana remain as the sole crops with little or no farm management activities (e.g. weeding, mulching or the removal of spent stems). Plantain has a greater nutrient demand and produces for only 2-4 years while banana can produce for 4-7 years. With proper attention with respect to weeding and clearing of undergrowth, this period can be extended to 10 years. In the Mbe Mountains complex, banana is cultivated more than plantain due to the low nutrient requirement and short period of maturity when compared with plantain.

Perennial Tree-crop Farms

Cocoa (*Theobroma cocoa*) farms are the major tree-crop farms in the villages of Okwangwo Division. Cocoa requires a deep loamy soil; hence small holder plots are often carefully chosen and can be found on level ground well inside the forest. Where the soil is considered sufficiently fertile, cocoa can equally be planted on slopes. It is a common practice to provide shade trees for cocoa. Such trees include the African pear (*Dacryodes edulis*) and occasionally cola tree (*Cola acuminata*) or avocado pear (*Persea americana*). These trees also provide economic yields (Holland et al., 1989). Once cocoa farms are established, the farm management tasks consist of weeding around the base of trees, application of pesticides and pruning. In most cases, these tasks are usually not completed satisfactorily, due to the lack of labour and other farm inputs including fungicides and pesticides. These constitute the major reason for the very low cocoa yields experienced by many households in the Division during the 1980s and 1990s. The trees appear to thrive under most conditions if provided with adequate shade in the early years. Thus, when well managed, cocoa could prove profitable especially to those in a position to purchase the necessary inputs.

6.4   Farming Practices

Traditional farming in Cross River State involves maintaining a bush fallow system around the village. In earlier times, it may well have been a true system of shifting agriculture. However, with increasing population, the introduction of new crops such as bananas and the availability of amenities such as roads and schools, the population have become more reluctant to move their villages as a shifting system would almost certainly require (Holland *et al.*, 1989). The bush fallow system practised in the Okwangwo Division involves the initial clearance of high forest (primary or mature secondary) for a mixed cropping of banana and plantain for a period of 3-10 years. This depends on the intensity of farm management activities undertaken to maintain soil fertility (e.g. weeding, clearing of undergrowth). This is followed by a fallow period of 3-5 years before the plot is put into one-year cassava cultivation, followed by another 3-5 years of fallow. Cassava continues to be cultivated after each fallow period until the households consider the soil to be infertile for further cultivation. Fallowed plots are rarely used for banana cultivation, as there is a tendency to clear fresh forest areas for the cultivation of banana.

A study of small farmers and forest loss in the Okwangwo Division (Ite, 1995; 1997) found that the length of fallow for many households (43%)

did not exceed five years while one-third of the households (34%) maintained that their fallow periods range from 6-10 years (Table 6.1). Some households (23%) claimed to leave their plots fallow for an extended period of 11-15 years. Fallow periods varied with the total number of plots per household and the crops to be cultivated. Although all farmers observed fallow periods, the number of farm plots per household and the crop to be planted after fallow determined to a large extent the duration of the fallow period. Farm plots with fallow periods of less than five years were normally utilised for staple food crop cultivation (e.g. cassava). A fallow period of 11-16 years implied that the plot was meant for banana and plantain cultivation. Fallow periods were extended if farmers had many plots but insufficient labour to cultivate their farm plots. From an ecological perspective, the study by Ite (1997) suggested that majority of the households had their farm plots under grass fallow, while one-third practice what could be described as the bush fallow system (see Bayliss-Smith, 1982). Only 23% of the households had their farm plots under forest fallow. Clearly, the length of fallow in the Okwangwo Division is not solely an agronomic decision, it also reflects the availability of labour (Ite, 1995).

Several factors determine farmers' choice of fallow regimes in the villages of Okwangwo Division. These include physical-ecological characteristics of the farm, ethnicity, length of residence and land tenure. Soils under forest vegetation are known for their inability to withstand annual cultivation without intensive management techniques (see Nye and Greenland, 1961). The fertility levels of the soil tend to be reduced quickly when cultivated as

Table 6.1 Length of Fallow Period by Households

Number of years	Abo Mkpang (N = 52)	Villages Kanyang 2 (N = 45)	Bokalum (N = 63)	Total households (N = 160)
<5	32.7% (17)	77.8% (35)	27.0% (17)	43.1% (69)
6-10	7.7% (4)	15.6% (7)	68.3% (43)	33.8% (54)
11-15	57.7% (30)	6.7% (3)	4.8% (3)	22.5% (36)
16-20	1.9% (1)	0.0% (0)	0.0% (0)	0.0% (0)

Source: Ite (1997)

leaching removes essential plant nutrients. This results in a farming system entailing the periodic movement from plots and the resulting fallow by farmers. Thus forest fallow is mostly found in areas further from the village where there are intensive land management practices while grass fallow are found closer to the villages.

Ethnicity and length of residence combined with land tenure influence fallow regimes (Chapter 5). For the majority of the indigenous (Boki) households, the number of farm plots under each fallow type is a quick indicator of the length residence of such households in the villages. It also provides further evidence about the land tenure arrangements internal to such households. Members of the community are apt to keep longer fallow periods. This is achieved by leaving their former plots and clearing new forest areas. Migrants are unable to maintain long fallow periods due to the land tenure arrangements (Chapter 5). Fallow periods of more than ten years are ecologically sound and allow soils to regenerate for cropping with efficient management and stable population (Gourou, 1980; Simmons, 1981). However, with population growth, greater farming pressures and shorter fallows, the bush fallow system is liable to break down leading to a loss of the ecological virility with the attendant problems of soil infertility and poor yields (c.f. Boserup, 1965). Overall, fallow periods in the Okwangwo Division has been gradually reduced in recent years due to population increase, the need for food crop production and the establishment of the Cross River National Park (Ite, 1997; Ite and Adams, 1998).

Although, the farming practice in the Division can be described as bush fallow system, households frequently clear fresh forest for plantain and banana cultivation. There are two possible explanations for this development. Firstly, clearing of forest entitles the clearer the perpetual right to use the plot; this in turn increases the land holdings of households. Secondly, it is the general belief among all the households that banana and plantain can not thrive well in plots other than those cleared from high forest or with at least 12 years of fallow to provide the necessary nutrients.

There is no doubt that notions of indiscriminate felling of trees by farmers in sub-Saharan African are highly exaggerated (Holmgren, 1994). In the study area, tree species are usually left in the farms during land preparation for cultivation, as admitted by four-fifths (86%) of respondents of a household survey (Ite, 1997). The most popular trees left by farmers include iroko (*Milicia excelsa*), black afara (*Terminalia ivorensis*), white afara (*Terminalia superba*), opepe (*Nauclea diderrichii*), bush mango (*Irvingia* spp.), camwood (*Pterocarpus osun*), mahogany (*Entandrophragma* spp.) and ebony (*Diospyras* spp.). Nearly all the respondents (97%) of the survey considered the trees to be economically significant either in the short or long-term. The trees left in farm plots also constitute important sources of building materials for domestic use as well as providing shade and protecting certain crops (e.g. banana and plantain) from wind damage. Trees such as *Pycnanthus angolensis, Ceiba pentandra, Musanga cecropioides, Brachystegia eurycoma* and *Parkia bicolor* have large leaves, which produce very good green manure, and are very often left, in the farms. The percentage of farmers practising clear felling of farm plots is relatively small. This clearly demonstrates the role and utility of indigenous agro-ecological knowledge in the management of forest and agricultural land in the study area.

The maintenance of soil fertility is crucial to smallholder agricultural practices in the Okwangwo Division. Households own farm plots in different locations and with various soil characteristics regarding types, and fertility levels. There are no significant differences among households in the villages with respect to their knowledge of the important role of soils in agricultural land use (Ite, 2000b). This illustrates the significance, understanding and utilisation of indigenous knowledge of the local physical environment by farmers in developing countries (Richards, 1985). As the Cross River Forestry Department (CRFD, 1994) observed, much detailed knowledge exists within communities in Cross River State concerning both natural and socio-economic environments. This covers the local characteristics of soils, flora, fauna, climate, processes and linkages. A spatial differentiation of household plots indicated that compound farms were cultivated with great intensity while plots adjacent or close to the compounds (e.g. cassava farms) are managed under an intensive fallow system (Ite, 1995). Furthermore, distant plots (e.g. banana farms) are put under an extensive bush fallow system. Under the bush fallow system, soil fertility is generally maintained by long periods of fallows (Nye and Greenland, 1961). The majority of the households in the villages of the Okwangwo Division do not use mulch or fertiliser to maintain the fertility levels of their farm plots. The main reasons relate to households' perceptions and understanding of soil fertility and their ability to afford fertiliser (Ite, 1995).

6.5   Constraints on Agricultural Production

There are four main constraints of agricultural production in the Okwangwo Division (Ite, 1995). First, there is insufficient supply of farm labour. The youths in the villages prefer hunting to farming; hence many are not directly engaged in full-time agriculture. Most households cannot afford the cost of hiring labour. Considering the extensive and intensive nature of the farming systems and the seasonality of farm operations, it is understandable why there are problems of labour supply. Yet, households would increase the size of their extensive operations rather than seek ways of intensifying production.

Second, the marketing of agricultural and other forest products in the study area is fraught with problems. However, there are differences between the villages in their perception and recognition of the problem. This can be attributed largely to the location and accessibility attributes of the villages. The villages located off the main road (i.e. the Ikom-Obudu Highway) experience significant difficulties in the marketing of their forest and agricultural products. This confirms the role of accessibility and transport as important indices in the development of markets as well as being a factor of forest loss in the Okwangwo Division (Chapter 7).

Third, pests and diseases constitute a constraint to agricultural development for the majority of the households in the Division. However, perceptions of the incidence of pests and disease tend to increase relative to the location of the village from the main road (Figures 4.2 and 4.3). One possible explanation is that villages along the main road are more exposed to the different techniques of pest and disease control and management than their counterparts in the remote villages. Agricultural pests and diseases are crop-specific. The major pests for cassava are cutting grass, porcupine and bush pig, which destroy the tubers. Mealybug (*Pseudococcus* spp.), leaf mosaic and tuber rot were widely mentioned as the major diseases of cassava. The major pests and diseases of banana and plantain include banana root weevil and fungal attack of the leaves and rhizomes. Black pod is the major disease of cocoa while capsids are the most destructive pests. Other common farm pests include elephants, bush pigs and porcupines.

Fourth, before the establishment of the Cross River National Park in 1991, farmer support services in Okwanwgo Division were non-existent (Ite, 1995). However, farmer support services in the whole of Cross River State was the statutory responsibility of the World Bank-funded and state-wide Cross River Agricultural Development Project (CRADP) launched in November 1985 (Etta, 1987). The project was designed to provide an integrated approach to improving the productivity of farmers, raise their incomes and improve the quality of life in rural areas. This was to be achieved through the stimulation of increased production of the country's basic staple food crops facilitated by training and visits (T&V) carried out by extension workers. The impact of the CRADP was not felt in the Okwangwo Division due to inadequate resources, although the CRADP ran a well-planned T&V extension service in Cross River State. Farmers in the Okwangwo Division faced several problems over inputs such as agro-chemicals, which were not locally available as required. This was the reason why the support zone development programme for the Okwangwo Division focused on agricultural development, especially through the provision of extension services (Chapters 4 and 8).

6.6   Conclusion

Agricultural production by households in Okwangwo Division is centred on cultivation in several plots, with each not more than 3 hectares. There is a gender bias (in favour of men) in the ownership of farm plots, and the labour force for their cultivation is relatively small compared to the needs of households. The current pattern of agricultural land use thrives on the extensive clearance of the forest for the cultivation of export or cash crops (e.g. banana), with shorter fallow cycles for food crops (e.g. cassava). Annual clearing of forest land by households for banana cultivation is motivated by the need to increase farm holdings and the local belief among the households that banana cannot thrive well except in farm plots made from freshly cleared forest areas. In the process of forest clearance for cultivation some trees are left in the farm, primarily for economic and also for environmental reasons. Chapter 7 examines the rates and implications of forest loss in Okwangwo Division against the background of the prevailing agricultural land use.

7 Rates and Implications of Forest Loss in Okwangwo Division

7.1 Introduction

This chapter explores the impact of agricultural land use on the forests in Okwangwo Division. This is achieved by examining changes in forest boundary prior to and after the establishment of the Okwangwo Division of the Cross River National Park. New data on forest extent, rates and patterns of loss in the villages of the Mbe Mountains complex are presented and discussed. Based on the mapping the forest extent and loss between 1967 and 1993, a spatial and temporal analysis of forest extent and loss in three detailed field study sites is undertaken. The underlying household socio-economic variables and cultural factors responsible for the observed forest loss are examined with a view is to articulating the role of small farmers in forest loss in the Division. The implications of the observed changes in forest boundary are discussed, with a focus on forest conservation and the Cross River State forestry sub-sector.

7.2 Forest Extent and Loss in Cross River State

The natural forest vegetation may have already been exhausted in some areas of Cross River State (CRS). As CRFD and ODA (1994) observed, a total of 15,270 km^2 of forest was lost during a 19-year period (1972-1991). Of this total, 91% was lost to farming and 9% to the development of plantations (see also Chapter 3). The total forest losses reported for the northern part of CRS was 23% (3,430 km^2), and in the southern and central parts of CRS, 51% (7,850 km^2) and 25% (3,990 km^2) of losses occurred respectively. Up to 36% (5,430 km^2) of the total losses in CRS were recorded in Forest Reserves while the remaining 64% (9,840 km^2) were in community-owned and managed forest areas.

Dunn *et al.* (1994) suggested that tropical high forest occupied 34% or 7,290 km^2 of the total land area of CRS (Table 7.1). A breakdown of this figure shows that 16% (or 3,307 km^2) of the tropical high forest in CRS were found within the Oban and Okwangwo Divisions of the Cross River National Park (see Figure 4.1). Less than one-fifth (19% or 3,983 km^2) of TMF were outside both Divisions of the park. According to Dunn *et al.* (1994), 6% (1,290 km^2) of TMF were in the areas described as the support zone of the National Park. Dunn *et al.* calculated that excluding the National Park, there was 1,810 km^2 of tropical high forest remaining within the forest reservation system in Cross River State. It seems that the total TMF extent of 34% reported by Dunn *et al.* were largely derived from the total area of Forest Reserves (including those within the Cross River National Park) and community forests in Cross River State.

Prior to the establishment of the National Park in 1991, there were 17 Forest Reserves in Cross River State, covering a total 6,101.29 km^2 or approximately 26% of the total land area of Cross River State (Ite, 1995). Although there were marked exploitation and illegal deforestation of the reserves, the total area of reserved forest land found in a series of Forestry Department annual reports (e.g. CRFD 1988, 1989, 1990, 1991, 1992) did not reflect these encroachments. These encroachments on the forest boundary took place after the reserves were originally surveyed and gazetted. For example, data collected by Osemeobo (1988) showed that the Ekinta Forest Reserve was close to total dereservation for food crop farming, yet the CRFD continued to include the reserve in its list. This contributes to the inflated figures of the actual extent of Forest reserves and indeed total forest extent in the State. As a result, the data by CRFD on the forest extent in Cross River State can be considered as misleading as this does not reflect the true situation on the ground. The remotely sensed data collected and analysed by Dunn *et al.* (1994) suggested that some Forest Reserves in Cross River State might

Table 7.1   Forest Type in Cross River State

Forest Type	Area (km^2)	Area (hectares)	%
Tropical high forest	7,290	729,000	34.3
Swamp forest	520	52,000	2.5
Mangrove	480	48,000	2.2
Oil palm plantation	219	21,900	1.0
Rubber plantation	146	14,600	0.7
Gmelina plantation	95	9,500	0.5
Other forests	216	21,600	1.0
Other land uses	12,299	1,229,900	57.8
Total	21,265	2,126,500	100.0

Source: Dunn *et al.* (1994).

have been cleared of vegetation. This supports earlier work by Okali (1989), who noted that as of 1981, up to 85% of the forest within the reserves were undisturbed, and that by 1989 (eight years later), less than 25% of some reserves could so be classified.

Data collected by Dunn *et al.*(1994) also indicated that in northern Cross River State, 7% of the total forest area lost during the period under review were in Forest Reserves while 93% were in community-own and managed forests. CRFD (1994) estimated that if this pattern of loss were sustained, the Forest Reserves in Cross River State could be completely cleared by the year 2014. With limited categories of vegetation, data presented in Table 7.1 reflects the degradation in much of the Forest Reserves of the Cross River State. It also provides the best possible estimate of TMF extent in Cross River State and, to an extent it sorts out the previous speculation on the extent and rates of forest loss in Cross River State (Ite, 1995). The differences in the estimates of forest extent in Cross River State presented in CRFD documents and data in Table 7.1 highlights the significant problems of measuring changes in forest extent. This relates mainly to the definition of forests and deforestation (see for example, WCMC, 1992), and the accuracy of estimates (Park, 1992; Grainger, 1993).

7.3   Forest Loss in the Mbe Mountains Complex, Okwangwo Division

Despite the scale of forest loss in Cross River State in southeast Nigeria, tropical forest is still the principal vegetation of the Okwanwgo Division. Ite and Adams (1998) have reported on a study of forest loss in three villages of the Mbe Mountains complex (Figure 7.1). Their findings strongly suggests that the rate of forest loss in the area has been very moderate (less than 1% per annum), especially when compared with estimates of national and regional rates of forest loss discussed earlier in Chapter 3. Table 7.2 presents data on the net change in forest extent during a 26-year period (1967-1993) for the three detailed survey villages.

From a spatial perspective, there are clearly defined patterns of the changing forest extent in the three study areas (Figures 7.2, 7.3 and 7.4). For example, it is clear from Figure 7.2 that prior to 1967, forest loss around the village of Abo Mkpang was concentrated along the western bank of the Okon River towards Bamba village in the north, rather than around the village itself.

Forest loss in Abo Obisu village (Figure 7.2) reflected the accessibility and relief of the area. As a result, forest clearance has now shifted eastwards towards Abo Bonabe community, and northeast towards the Okon River, in a

Table 7.2   Net Change in Forest in Mbe Mountains Complex, 1967-1993

Study Area	Total area mapped (km²)	Forest area (km²)		Change in extent 1967-1993 (km²)	Annual loss 1967-1993(%)
		1967	1993		
A	92	82.5	72.5	10	0.47
B	49	43.0	40.0	3	0.27
C	72	61.5	49.5	12	0.75

Source: Ite and Adams (1998)

bid to avoid the steep slopes. The northward and southward orientation in the forest clearance patterns suggests that the Ikom-Obudu Highway probably served as a catalyst in the process of forest loss to agriculture in the village.

In the second survey area (Figure 7.3), forest clearance prior to 1967 had focused on the village of Kanyang 1 and along the Ikom-Obudu highway. Kanyang 2 is a recent settlement established more than 50 years ago, when some family groups broke away from the original Kanyang (now Kanyang 1). The new settlers from Kanyang 1 undertook the initial clearance of forest around the new village. Between 1967and 1993 (a period of 26 years), there was an eastward shift in forest clearance, from the eastern banks of the Afi River across the Ikom-Obudu highway towards the Mbe Mountains conservation and tourism zone (Figure 7.3).

On the other hand, the spatial pattern of forest clearance in the third study area (around Bokalum village) is different from the previous two areas (see Figure 7.4). Forest clearance before 1967 had been concentrated around Bokalum village. It has been suggested (Ite and Adams, 1998) that the difficult terrain of the Mbe Mountains may have prevented further forest clearance to the southwest of Bokalum village. As such, forest loss in the village after 1967 was linear in shape, especially along the feeder road linking Wula 2 to Bamba and towards Butatong village in the north (Figure 7.4).

From Table 7.3, it is clear that 7% of land in the three study areas were already under some form of permanent cultivation in 1967. Some 5% had been under fallow since 1967 and allowed to return to closed-canopy forest. A further 12% was cleared between 1967 and 1993. By 1993, about three-quarters (76%) of the forest in the three study areas remained untouched (Table 7.3). The data suggest an average loss of 4.17 km² of forest area per village within 26 years (1967-1993). It is estimated that 38 km² (3,800 ha) of forest would have been lost in the Mbe Mountains complex between 1967 and 1993, and approximately 150 km² (15,000 ha) for the whole of the

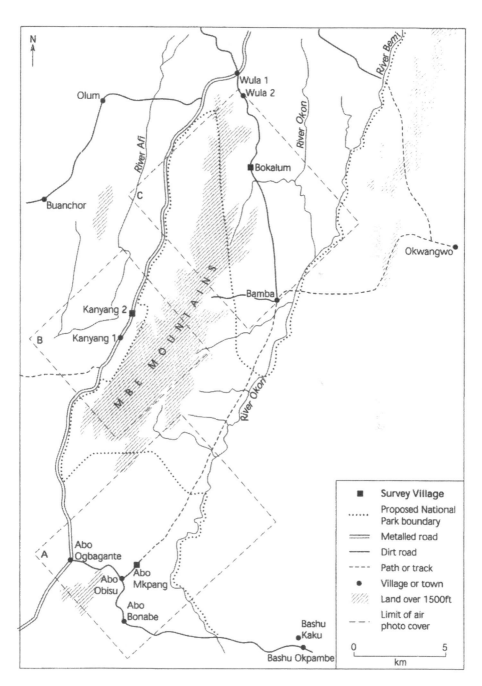

Figure 7.1 Detailed Study Sites

SITE A

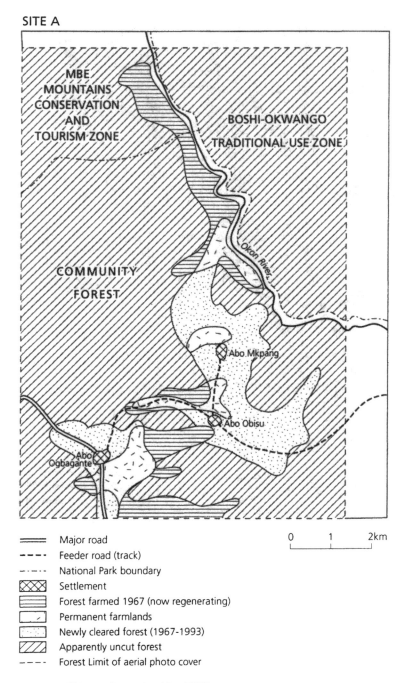

═══	Major road
‒ ‒ ‒ ‒	Feeder road (track)
‒ ∙ ‒ ∙ ‒	National Park boundary
⬡	Settlement
▤	Forest farmed 1967 (now regenerating)
▢	Permanent farmlands
▦	Newly cleared forest (1967-1993)
▨	Apparently uncut forest
‒ ‒ ‒ ‒	Forest Limit of aerial photo cover

0   1   2km

Figure 7.2   Forest Loss in Abo Villages

SITE B

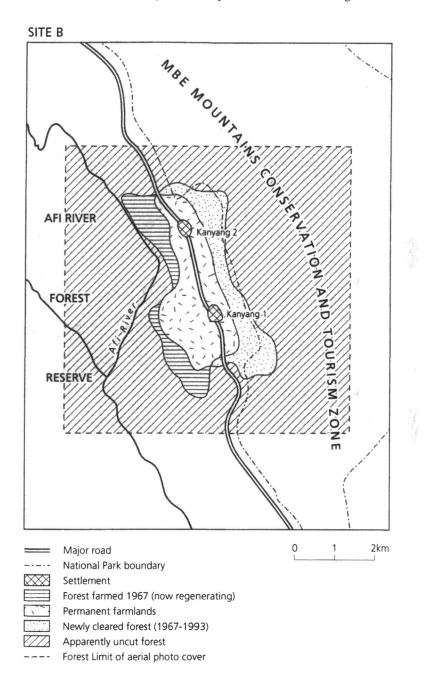

Major road
National Park boundary
Settlement
Forest farmed 1967 (now regenerating)
Permanent farmlands
Newly cleared forest (1967-1993)
Apparently uncut forest
Forest Limit of aerial photo cover

0   1   2km

Figure 7.3   Forest Loss in Kanyang Villages

SITE C

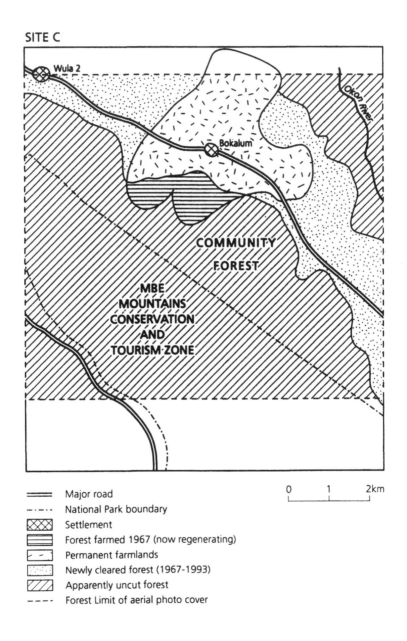

Major road	
National Park boundary	
Settlement	
Forest farmed 1967 (now regenerating)	
Permanent farmlands	
Newly cleared forest (1967-1993)	
Apparently uncut forest	
Forest Limit of aerial photo cover	

Figure 7.4 Forest Loss in Bokalum

Table 7.3   Remaining Forest Extent in Mbe Mountains Complex

Survey Area	Total area mapped, km²	Area under permanent cultivation before 1967, km²	Area cultivated 1967, reforested 1993, km²	Area forested 1967, cleared 1993, km²	Apparently uncut forest 1993, km²
A	92 (43%)	3.75 (4%)	5.75 (6%)	10 (11%)	72.5 (79%)
B	49 (23%)	45 (9%)	1.5 (3%)	3 (6%)	40 (82%)
C	72 (33%)	8 (11%)	2.5 (3%)	12 (17%)	49.5 (69%)
Total	213 (100%)	16.25 (7%)	9.75 (5%)	25 (12%)	162 (76%)

Source: Ite and Adams (1998)

Okwangwo Division during the same period under review (Ite and Adams, 1998). The total forest loss of 150 km² in the whole of the Okwangwo Division during the period is equivalent to an annual loss of 6 km² or 0.6%. This calculated extent of forest loss in the Okwangwo Division therefore provides strong empirical support to the work of Dunn *et al.* (1994). The survey by Dunn *et al.* (1994) had concluded that only 20% of the TMF of the Division had been cleared, and that forest loss in the forest reserves constituting the National Park was very limited.

7.4   Role of Small Farmers in Forest Loss

The role of small farmers in forest loss in the Okwangwo Division, particularly in the Mbe Mountains villages can be examined from seven main perspectives. These are:

- Primacy of banana cultivation:
- Ethnicity of households:
- Impact of migration:
- Accessibility of the villages:
- Educational levels and employment opportunities:
- Institutional arrangements for forest resource management, and
- Reality of and aspirations for socio-economic development.

Primacy of Banana Cultivation
Prior to the establishment of the Cross River National Park, the villages of

Okwanwgo Division were known mainly for the cultivation of cocoa, cassava, yams, water-yams, potatoes, plantain, and oranges, both for subsistence and for sale. Cocoa was the paramount cash crop prior to the early 1970s. This evident in the total number of such farms, many of which became abandoned for various reasons including, the high demand for labour, problems of pests and diseases and the grading and pricing systems. Farmers felt that cocoa cultivation was no longer economically viable under the prevailing economic climate (Ite, 1997). Bananas have now replaced cocoa as the main cash crop. The increasing demand for bananas both within and outside Cross River State led to their widespread cultivation, and the resulting boom in the trading in bananas. The decision by households to transfer their farm resources from cocoa cultivation to banana farming are influenced by several considerations (Ite, 1995). These include the fact that bananas are less prone to pests and diseases; demand less and cheap labour; have a shorter period of maturity; and are easy to harvest as well as being used for domestic consumption.

Since 1990, trading in bananas has been highly organised in the villages Okwangwo Division, especially those around the Mbe Mountains complex. Potential buyers have been required to register annually with the village council. Many households sell their bananas to resident registered buyers or their agents (some of whom are indigenes of the village), while others deal directly with visiting buyers, mostly from ethnic groups outside Cross River State, especially the Ibos. The quantity of bananas available in each village on a particular market day determines to a large extent, the price buyers would be willing to pay for a bunch of banana. As Ite (1997) observed during fieldwork in the villages, most buyers have capitalised on the fact that bananas are perishable and that due to huge transport costs, the households themselves cannot transport bananas from the villages to the nearest major market centres outside the village. It is common for households to sell at 'give away' prices to avoid waste. As such, the marketing arrangement for bananas can be considered as unfavourable to the farmers. Buyers determine the price paid and not the farmers. The limiting factors in the successful undertaking of banana trade by local households include the availability of credit for the hiring of lorries, labour, and actual purchase of bananas from the villages.

Ethnicity of Households
Ethnicity is a significant inherent asset for households in the Okwangwo Division, as it determines access to land (Chapters 2 and 5). The Boki people of northern Cross River State mainly inhabit the Division. Much of the community land and forest area, except the Forest Reserves, is held under customary laws and traditions. Since there are no legal titles, ownership of land is largely communal. Indigenous households from the respective villages have open

access to the community forest. They are allowed to clear as much forest as possible for the purposes of farming. The form of communal ownership of land has implications for land use and soil fertility management (Chapters 5 and 6).

The fact that individual ownership of land derives from the initial clearing of forest encourages the clearance of more forest, sometimes not accompanied by any form of cultivation. On the other hand, the existing land tenure system does not provide much room for the indiscriminate clearance of forest vegetation by non-indigenes. Migrants are not allowed to clear fresh (primary) forest areas, but could lease secondary forest areas from indigenous households, who in turn would be free to clear more primary forest areas. Although it has been argued elsewhere in this chapter that the arrival of migrants in the villages contributes to forest loss in the study area, the scale of such losses can be considered low when compared to those of indigenous households.

Impact of Migration
When compared with other states in southeast Nigeria, there is no doubt that extensive areas of apparently uncut forest still exist in Cross River State (see also Caldecott and Morakinyo, 1996). The existence of forests around most communities in Nigeria tends to be interpreted as a sign of significant socio-economic underdevelopment (Atte, 1994). As a result, such places might be perceived as unattractive for the purposes of settlement. However, Ite (1997) reported that within a five-year period, the nine villages of the Mbe Mountains complex experienced a 7% increase in the total number of households, as a result of in-migration of both indigenes and non-indigenes. Migrants from at least five states of eastern Nigeria have been recorded (Ite, 1995). These include the Ibos (from Abia, Imo, Anambra and Enugu States), the Ibibios (from Akwa Ibom State), the Obanlikwu and Obudu people (from northern Cross River State), and non-Nigerians (e.g. Ghanaians). The majority of migrants are engaged in economic activities dependent on forest exploitation especially within the agricultural sector of the village economy. Most are actively engaged in the cultivation and marketing of bananas and plantains, known to be largely responsible for annual clearance of large extent of fresh forests.

It is important to note that the source regions of the first two groups of migrants above are characterised by high population densities and people-land ratios as well as an acute shortage of farmland (Udo, 1990), hence their decision to migrate to areas with sufficient agricultural land. Households and communities justified on economic grounds, the acceptance of non-Boki migrants into their villages. As a result, indigenes have been motivated to leave old farms for the migrants and to proceed to the forest frontiers to make fresh clearing of forest areas. The continued acceptance of migrants by households and forest communities has clear implications for an increase in absolute population figures, and increasing demand for cultivable land. It is deemed that the pressure to accept migrants in the villages of the Park area can only increase in future along with forest loss (Ite, 1997; Ite and Adams, 1998).

Accessibility of the Villages

The importance of accessibility and transportation in the socio-economic development process cannot be over-stressed. The location and accessibility of villages influences households' decisions on agricultural production (see also Veldkamp *et al.*, 1992). Within the context of the Okwangwo Division, a good case in point is the frequency and price at which harvested bananas are bought, sold and transported from the villages. Although it has not been statistically proven, it is evident that there is a significant relationship between the location and accessibility of the villages and the income derived from farm produce, especially bananas (Ite, 1997). Farm incomes in villages with the benefit of access roads and transport were reported to be higher than those in villages served with inadequate transport infrastructure (Ite, 1995). This disparity in the price system stems from the mode and cost of transportation of bananas from villages along the secondary feeder roads to the major centres along the Ikom-Obudu highway.

In adapting to this perception, several households in the remote villages of the Okwangwo Division adopt three main strategies. The first is to increase the number of farm plots devoted to banana cultivation; the second is to expand the size of present farms; and the third is to transport bananas from the farms to as near the highway as possible. However, the effectiveness of these adaptation mechanisms has been limited by several factors including the distance from the farms, head-loading rates, and the availability of labour (Chapter 6). It can be argued that poor facilities for marketing bananas have therefore limited the capacity for the alleviation of poverty in the villages of the Mbe Mountains complex through increased agricultural production. On the other hand, the Ikom-Obudu Highway can be seen as providing an added incentive for more indigenous smallholder banana cultivation in the villages (Ite, 1997). Prior to the opening of the road, banana cultivation and trade were less significant in household economics. With the road in place, the villages have now been linked to the wider market system for this agricultural produce. However, it is unclear whether forest loss in the area could be checked by improved the marketing conditions for banana that favoured farmers.

Educational Levels and Employment Opportunities
Education opens up the possibility of employment opportunities that are not necessarily dependent on the exclusive use of forest and land resources. The literacy rate in the communities of the Division is low when compared to other parts of the Cross River State. Primary data collected by Ite (1995) showed that 48% of the household heads had primary education as the highest level of educational attainment while one-quarter (24%) have had secondary education. Furthermore, the data on the educational level of adult household members revealed that at least 25% of the households had two adults with no formal education, while 53% had at least two members with primary education. There is no doubt that educational levels determine occupational choice. Household heads are emphatic on their desire for the best possible education for their children to enhance their chances of better and well-paid jobs outside the forest communities of the Division (Ite, 1997). Yet, in the area, there are few schools (Chapter 5), and in some cases, children have had to board in schools outside their villages. This requires substantial financial implications, which in this case has to be derived from the exploitation of available natural resources - the forest. As such, economic circumstances resulting from inadequate education tends to increase pressure on forests through conversion to farmland and adoption of the agricultural practices and associated cropping systems (Chapter 6).

Local Institutional Arrangements for Forest Resource Management
The traditional forest resource administration and management systems in the villages of the Okwangwo Division possess decision-making processes, rules and regulations on resource use, and patterns of resource ownership as well as in-built social regulatory mechanisms about the use of natural resources (Chapter 5). However, it appears that such institutions, particularly the land tenure system are increasingly unable to exert sufficient control over the exploitation of the forest resource. Sanctions for the abuse and misuse of forest resources have not been as effective as they were three decades ago. Some community members disregard traditional laws and customs in the processes of forest exploitation for agricultural purposes and are not sufficiently punished. The breakdown of traditional systems of land resource management has been attributed to inefficient village administration and general household poverty (Ite, 1995). In recent times, the need for money has become the overriding concern for many households. Household heads are now more preoccupied with the quest to meet the bare essentials of food, shelter, clothing to maintain minimum levels of living.

On the other hand, it can be argued that the establishment of the National Park has been a contributory factor in the rapid rate of forest loss in

Okwangwo Division (Ite 1997; Ite and Adams, 1998). Prior to this arrangement, the protection and management of the forest estates of Cross River
State was the statutory responsibility of the Cross River Forestry Department
(CRFD). The inability of the CRFD to effectively carry out its assigned role
led to a series of encroachments into reserved forest areas in the CRS. In the
Okwangwo Division, local communities are fully aware of the negative implications of the National Park project, especially in relation to the future
availability of forest areas for farming (Chapter 8). This resulted in an increase in the rates of forest clearance prior to the full enforcement of National
Park regulations (Ite, 1996). For example, approximately 12% of the total
number of farms in the villages of the Mbe Mountain were made by households four years after the establishment of the Cross River National Park in
1991 (Ite, 1997).

Socio-economic Development: Reality and Aspirations
The forest is regarded as a valuable natural resource among the rural popu-
lace of Cross River State. Communities acknowledge the important role of
the forest area and its resources in providing agricultural land, vital medicinal
plants, building materials, food, source of income, protecting watersheds,
among other socio-economic, cultural and environmental values (Ite, 1995).
On the other hand, the availability of money is widely recognised as a measure of development in the villages around the Division. From the household
perspective, poverty is defined in terms of the ability of the household head to
feed, house, clothe and bring up offspring according to the dictates and re-
quirements of the modern social economy. It also includes the availability of
capital to expand farms, hire labour and the credit facilities to maintain them.
These in turn are influenced by the availability or otherwise of socio-economic development infrastructure at the community level (Chapter 5). Thus
at the household level, it becomes essential to increase crop production and to
generate more income for the purposes of improved living standards (Chapter 2), using the income opportunities available to each household (Osemeobo,
1993b). The net result in Okwangwo Division has been the gradual and cumulative forest loss due to the need to turn into reality, household aspirations
concerning their socio-economic development, including education and better health care.

**7.5   Changes in Forest Extent: Implications for Conservation and
         Forestry**

The observed changes in the forest extent of Okwangwo Division has several implications, two of which are particularly relevant to the arguments presented in this book.

Implications for Conservation Policy Response
The establishment of the Cross River National Park project in 1991 was a direct and significance response to speculative notions and perceptions of very high rates of forest and biodiversity loss in Nigeria and Cross River State (Caldecott, 1996). However, these concerns were not wholly borne out or substantiated by socio-economic and ecological research in the Okwangwo Division. There was increasing fear that accelerated forest fragmentation due to the prevailing agricultural practices (Chapter 6), would eventually result in widespread and extensive insularisation of wildlife (especially gorilla) habitats, and that present rates of forest loss could degenerate and trigger possible species loss and extinction in the Okwangwo Division.

As a result, the possibility of maintaining a pristine forest connection between the Afi River Forest Reserve and the forests of the Mbe Mountains complex as a wildlife corridor was explored during the design and zoning of the Park area (see Shuerholz *et al.*, 1990). As Ite and Adams (1998) noted, wildlife biologists began developing the theory of wildlife corridors in the mid-1970s, with the idea anchored on the island theory of biodiversity (Shafer, 1990). It was adopted in the early 1980s and promoted by such influential international conservation organisations such as the World Wide Fund for Nature and the World Conservation Union. In sub-Saharan Africa, wildlife corridors have been suggested in many places, for example on the land area linking the Maasai Mara Reserve (Kenya) with the Serengeti (Tanzania) as well as the East Usambara Mountains and on Mount Kilimanjaro in Tanzania (Newmark, 1993).

For the Okwangwo Division, Oates *et al.* (1990) suggested that such a corridor could allow gene flow between wildlife populations in the Afi River forest reserve and the Mbe Mountains complex (Figures 4.2 and 4.3). However, it is hard to find practical evidence of the success of these corridors in practice in environments such as those of the Cross River National Park.

From a purely conservation perspective, the low rates of forest loss within and around the Okwangwo Division are significant, and demand a policy response from the national park management. The study by Ite (1995, 1996) found that households in the surrounding villages continued to exploit the forest resources within the park, especially the forest areas of the Mbe Mountains conservation and tourism zone. This was attributed partly to low level of park management operations at the time and a demonstration of resentment at lack of visible socio-economic benefits from the development of the Park (Chapter 8). The communities had a justifiable feeling of neglect and abandonment by the park project, which had promised several socio-economic incentives (Chapter 4), in return for their acceptance of the annexation of the Mbe Mountains complex to the Park area.

Implications for Forestry in Cross River State

The resources of the forest estate of the Cross River State represent one of the main assets for its economy (Dunn *et al.*, 1994). In 1994, the estimated total tariff value of the total volume of timber (standing crop) in the tropical high forest (excluding the Cross River National Park) was estimated at N5.6 billion (CRFD, 1994). In 1993, the officially recorded annual volume of timber harvested in Cross River State was 108,000 m^3. In 1993, an estimated 64,000 m^3 of wood (worth N191 million) was exported to other states in Nigeria. Furthermore, in the same year the State government was reported to have derived N2.8 million from timber tariff (CRFD, 1994), against N2.6 million in 1991 (CRFD, 1991).

In Cross River State, both the large timber processors and small-scale power chain-saw operators harvested timber from the forest as at 1994. Logging was a minor economic activity in the Okwangwo Division (see also Dunn *et al.*, 1994). There were no large-scale timber concessions. Single tree permits were issued to power chain-saw operators who were the major exploiters of timber from the forests of the Division. Nonetheless, there were cases of operators exploiting timber without proper legal authorisation from the Cross River Forestry Department (CRFD). Although they were selective regarding the species removed, unsupervised logging damaged other trees and created opportunities for expansion of farms by small farmers (Ite, 1995).

Despite the fact that it constituted an offence for operators to engage in timber exploitation without appropriate license, there were only 541 registered power chain-saw operators in Cross River State, with an estimated 1,500 unregistered nor their activities regulated (CRFD, 1994). The higher charges levied by the CRFD encouraged illegal logging by the small-scale operators. This was compounded by the inability of the CRFD to control such activities, resulting in more revenue being lost than collected from timber exploitation. Overall, the changes in the forest extent of Okwangwo Division have implications for the timber prospects in Cross River State in general and the revenue generating capacity of the CRFD in particular. However, it is pertinent to note that in spite of the fact that the CRFD was a major revenue source for the state coffers, there is evidence to show that it was always under-resourced for effective forest resource management operations (Okali, 1989; Armstrong *et al.*, 1990; Atte, 1994).

7.6   Conclusion

Evidence presented in this chapter provides an indication of the rates of forest loss in the Mbe Mountains complex in particular and the Okwangwo Division of the Cross River National Park in general. Each of the three detailed study sites exhibits unique spatial characteristics, and provide a much clearer picture of forest extent and loss in the area prior to and after the establishment of the Cross River National Park in 1991. It is evident that forest loss in the study area fit into the general pattern of decreasing forest extent in Cross River State, Nigeria and indeed West Africa. The role of small farmers in the observed patterns of forest loss in Okwangwo Division has been examined from the socio-economic and cultural context of the households in the study area. Political economic and human ecological processes operating within and outside the villages underlie the explanation of forest loss in the Division.

8 Forest Conservation-with-Development in Okwangwo Division

8.1 Introduction

This chapter evaluates the attempt to conserve the forests of the Okwangwo Division of the Cross River National Park, based on the principles of conservation-with-development (Chapters 1 and 4). It reviews and evaluates the practice of conservation-with-development, especially by examining the pilot phase in the Mbe Mountains conservation and tourism zone (1991-1994) and the European Union-funded Okwangwo Programme (1994-1998). The objective is to explore the extent to which the design and implementation of such initiatives effectively stabilised and checked forest loss, especially in the Okwangwo Division where there is a good history of local resource use and management.

8.2 The Pilot Phase in Mbe Mountains Villages (1991-1994)

Two separate but inter-related projects were initiated in 1991 in the Okwangwo Division of the Cross River National Park. These are the Mbe Mountains Conservation Project and the Mbe Mountains Support Zone Development Programme. These initiatives served as the pilot phase of the implementation of integrated conservation and development project (ICDP) activities as reflected in the 1990 management plan for the Division (Caldecott *et al.*, 1990a).

The Mbe Mountains Conservation Project (MMCP) was established in July 1991 based on evidence of the poaching of gorillas within the Mbe Mountains complex. With major funding from Shell Petroleum Development Company of Nigeria, the MMCP formally commenced with the recruitment of nine Park Rangers. It had the goal of protecting the gorillas in the Mbe Mountains complex as well as consolidating the activities of the Gorilla Conservation Project (GCP) at Kanyang. The GCP was initiated in 1988 by the Nigerian Conservation Foundation, pending the amendment of the definition of the park boundaries and the sourcing of funds for the full implementation of the Cross River National Park project (Chapter 4). Through its team of enforcement staff, the MMCP effectively controlled logging and hunting activities in the core area of the Mbe Mountains complex, a conservation and tourism zone of the Okwangwo Division. However, the protection of the Okwangwo Division in general was fairly low-key based on the understanding by the project management and other interest groups that a new National Park cannot exist in an atmosphere of confrontation and grievance.

Stricter enforcement of National Park laws and regulations (i.e. based on Decree 36 of 1991) was expected to begin after the receipt of funding from the European Community for the protection and development programmes of the National Park (Marshall, 1993b). However, the patrolling of the Mbe Mountains complex by the staff of MMCP became more difficult during the last quarter of 1993. This was due to the hostility of some villagers over the decision by the management of the CRNP project to site the Okwangwo Divisional headquarters at Butatong. Butatong village was not included as a beneficiary of the Mbe Mountains Support Zone Development Programme framework (described below), although it was located within the support zone of the Okwangwo Division. During this period, local people denied Park Rangers and some tourists access to the Mountains and the Park area. As a result, incidences of poaching (e.g. elephants, drill and chimpanzee) and illegal extraction of forest products flourished without any hindrance. In its totality, the MMCP provided an early learning experience of park protection, which was to be applied later in the management of the entire Okwangwo Division.

On the other hand, the Mbe Mountains Support Zone Development Programme (MMSZDP) was set up in August 1991, with a two-year funding from the Ford Foundation. It complemented the MMCP and both formed the early forest conservation project concept for the Okwangwo Division (Ashton-Jones, 1992). The MMSZDP acted as an interim measure of indirect compensation to the nine villages surrounding the Mbe Mountains complex for their loss of access to the area designated as the National Park (Usani, 1992a, 1992b). The immediate purpose of the programme was to ensure that the Mbe Mountains villages fully benefited from the income generating activities arising from the support zone development programme of the National Park (Chapter 4). The long-term goal of the MMSZDP was to involve the people at the early stages of Park development so that they would eventually wish to conserve it themselves. The overall SZDP of the Cross River National Park project placed considerable emphasis on an effective agricultural extension service as the key developmental operation in the support zones of the Park (Chapters 4 and 6). The agricultural development component of the MMSZDP consisted of two main elements viz.: the administration of loans from a revolving credit fund scheme and the provision of farmer support services in the villages of the Mbe Mountains complex.

The revolving credit fund scheme moved from the drawing board to the field in March 1992, when the first pre-loan village development workshop was held at Abo Mkpang village (Ashton-Jones, 1992). The aim of the workshop was to provide the nine villages of the Mbe Mountains complex with detailed information concerning the loans on offer as part of the development activities of the MMSZDP. In disbursing the loans, priority was given to community groups who were interested in food crop farming activities as opposed to cash or export crop production. For example, the MMSZDP attached greater importance to cassava cultivation, and this was justified as an attempt to increase food sufficiency at the household level, and to produce a surplus for sale. This should, in turn, contribute to subsequent increase in household incomes and wealth. The underlying idea was that farmers would possibly consider intensive farming systems and hence reduce dependence on annual clearance of fresh forest areas. However, food self-sufficiency was not a development problem in the communities of the Mbe Mountains complex (Environment and Development Group, 1998). Although the majority of households were aware of the offer and mode of operations of the loans, less than 20% had derived any direct benefit from their disbursement and administration. Overall, the loans had a much more limited impact on households than was envisaged by the management of the MMSZDP (Ite, 1995).

The organisation and provision of farmer support services in the Mbe Mountains villages was based on the same principle as the World Bank's unified 'training and visit' extension system described in Chapter 6. The MMSZDP extension agents were assigned several responsibilities, including teaching and demonstrating skills to farmers, motivating farmers to adopt production recommendations, and bringing farmers' agricultural production problems to the attention of the subject matter specialists. However, by the end of 1993, there was clearly very little extension work in the villages for several reasons (Ite, 1995). For example, the extension agents were not properly equipped with the necessary resources. Most of the farm implements supplied by MMSZDP were inappropriate and irrelevant. In addition, the formal introduction of the extension agents to the communities only began in the last quarter of 1993, but with significant problems arising from the cantankerous nature of the relationship among the Mbe Mountains villages, especially in relation to the derivation of benefits from the community forest area.

In retrospect, the MMCP and the MMSZDP constituted very useful institutional development and training exercise for the forest conservation project in Okwangwo Division, as much as it was a development project for the nine villages of the Mbe Mountains (Marshall, 1993b; 1994a; 1994b; Hurst and Thomson, 1994). In the process, the conservation project management realised that such work could not be rushed, as it required time and patience on the part of both the developers (the park project) and the developed (local communities).

8.3   Initial Community Perceptions, Expectations and Support

An evaluation of Mbe Mountains communities' perceptions, expectations and support of the MMCP and MMSZDP found that just over one-third of the households had derived benefits from the pilot phase of the integrated conservation and development project in the Okwangwo Division (Ite, 1995; Ite and Adams, 1998). There was strong evidence of a significantly low level of community support for the Park project in the villages of the Mbe Mountains complex. From the development perspective, the MMSZDP was an attempt to stabilise agricultural land use around the protected area by addressing the socio-economic needs of households and communities. Overall the short-term results were less impressive (Ite, 1996a; Ite and Adams, 2000), which is typical of such projects in sub-Saharan Africa (Hannah, 1992; Wells *et al.*, 1992; Barrett and Arcese, 1995). Few households in the villages of the Mbe Mountains complex appreciated the benefits of the National Park. Benefits such as employment, tourism, education, medical services and transport only accrued to a marginal percentage of the local population (Table 8.1), while the threats to the integrity of the park emanated from beyond the boundaries of the national park and the support zone villages.

The majority of households from the Mbe Mountains communities were dissatisfied with the immediate (and prospective long-term) disadvantages of the park. These included the reduction in agricultural land area as well as

Table 8.1   Perceived Benefits of National Park (ranked)

		Villages		
Benefits	*Abo Mkpang*	*Kanyang 2*	*Bokalum*	*Total*
	(N = 31)	*(N = 7)*	*(N = 17)*	*(N = 55)*
Employment	58% (18)	43% (3)	65% (11)	58% (32)
Tourism	84% (26)	43% (3)	18% (3)	58% (32)
Environmental education	26% (8)	14% (1)	29% (5)	31% (17)
Health care	36% (11)	14% (1)	35% (6)	27% (15)
Transport	7% (2)	0% (0)	12% (2)	24% (13)

Source: Ite (1996)

losses of traditional hunting areas, fishing rights, gathering areas and access to economic trees (Table 8.2). Where households and communities supported the forest conservation project, it was because they expected benefits that were significantly different from those planned by the National Park. Households and communities that derived any form of benefits from the park regarded the MMSZDP as a development agency, whereas those whose traditional resource use patterns are disrupted regarded the MMCP as promoting only the conservation of nature, desiring the area devoid of all human occupation and exploitation. However, it has to pointed out here that the evidence of insufficient support for the National Park by households in the villages of the Mbe Mountains complex does not necessarily mean general ignorance of the value of conservation on the part such communities in developing countries (Harcourt *et al.*, 1986). Research by Ite (1996a) found an overwhelming community awareness of the need for forest conservation in the Okwangwo Division.

Several variables influenced the level of initial support that was extended to the forest conservation project by local communities of the Mbe Mountains complex. These include ineffective macro-economic planning and management, regional and local under-development, limited external conservation project funding and timing, poor public relations strategies, disregard and ignorance of local historical and political aspects of local resource control, especially during project planning and implementation.

Table 8.2   Perceived Disadvantages of the National Park (ranked)

Disadvantages	Abo Mkpang (N = 21)	Villages Kanyang 2 (N = 38)	Bokalum (N = 46)	Total (N = 105)
Reduced agricultural land	81% (17)	87% (33)	74% (34)	80% (84)
Loss of hunting area	62% (13)	76% (29)	87% (40)	78% (82)
Loss of fishing rights	67% (14)	42% (16)	74% (34)	61% (64)
Loss of gathering area	62% (13)	13% (5)	94% (43)	58% (61)
Loss of access to economic trees	43% (9)	42% (16)	78% (36)	58% (61)

Source: Ite (1996)

8.4   Wider Institutional Issues: Co-operation and Conflict

During the planning of the Cross River National Park (CRNP), Holland *et al.*, (1989) found that the work programmes of federal government and Cross River State institutions were relevant to general rural development of the support zone villages of the CRNP. Such work programmes included the provision of agricultural extension and associated technical services, construction and maintenance of rural feeder roads as well as the provision of health and education facilities. In most cases, there was considerable overlap between the work programmes of these agencies and their subsidiary units. However, research evidence (Ite, 1995, 1998) strongly suggested that initially, the Okwangwo Division (as an institution) was quite isolated from many of the federal and state institutions and organisations identified by Holland *et al.* (1989). In the few cases where any relationship and support existed, they were not solid or built on complete trust and understanding. A good example is the relationship between the CRNP and Cross River Forestry Department (CRFD). Several institutional conflicts arose when reserved forest areas were transferred from the CRFD to the CRNP.

First, the CRFD considered the creation of the CRNP as a significant reduction in the forest areas under its management, and hence a reduction in their revenue generating capacity. Prior to the establishment of the CRNP, the CRFD was more of a revenue generation agency than a forest management organisation (Chapter 7). As an arm of the Cross River State civil service, the CRFD made every effort to maintain its *status quo* and professional integrity, even in the face of declining productivity resulting from several problems (Atte, 1994). The creation of the CRNP meant a possible reduction in the Forestry staff strength due to reduced forest area under the management of the CRFD, despite the fact that both Divisions of the CRNP constitute approximately 21% of the total forest estate of Cross River State (Ite, 1995).

Second, there were cases of duplication of forest management and rural development efforts in the support zone villages by the CRNP and the CRFD. It was not always very clear which of the two institutions had the sole statutory responsibility of enforcing laws and regulations concerning poaching or illegal timber extraction within the forest areas constituting the national park. The CRFD laid its claims as the primary manager of forest estates of Cross River State and sought to continue with this role. On the other hand, the CRNP maintained that Decree No. 36 of 1991 (National Parks Decree) superseded the Forest Reserve settlement orders and forest laws of the 1930s, which formed the basis of forest conservation and management by CRFD. A case of open confrontation between the staff of CRNP and CRFD was recorded in Oban Division of the National Park.

Third, conflict arose between and CRFD and the CRNP concerning the definition, size and specific management objectives of the support zone of the national park. As CRFD and ODA (1994, p.13) put it:

> ...to-date, discussions between FDD and the CRNP Authority have been inconclusive concerning the size of the Support Zone and the multi-sectoral arrangements for planning and managing the Zone's natural, economic and social resources in an integrated manner. The rationale for the size of Support Zone currently designated is far from clear.

> The FDD strongly advocated that the Support Zone should constitute the whole of the area of the State outside the National Park. In its view, to do otherwise would most probably result in resource allocations, which would in the medium to loner-term prove to be counter-productive.

Overall, by 1996, the entire CRNP project was considered to be under significant threat from the forest resource exploitation policies of the both Cross River State and the Federal Government of Nigeria. For example, the former planned (with the support of the latter) to grant extensive logging concessions to a Hong Kong-based Chinese company, Western Metal Products Company. However, this decision met strong opposition from various national and international pressure groups and environmental organisations. Many of these organisations are well known for their keen interest in and significant action towards environmental and resource conservation in Nigeria (Chapter 3).

8.5   The EC-Funded Okwangwo Programme (1994-1998)

The Okwangwo Programme was the European Community's four-year programme of investment in the Okwangwo Division of the Cross River National Park, based on the revised management plan and the 1993 Financial Agreement for the Okwangwo Division. The Programme commenced in July 1994 with a budget of ECU 4.087 million, with additional support for technical assistance from WWF-UK's Joint Funding Scheme (Environment and Development Group, 1998). The total budgetary allocation for the Okwangwo Programme was less than 25% of the funding envisaged in the original (i.e. 1990) management plan for developing the Okwangwo Division (Chapter 4). The Okwangwo Programme had several wider objectives, including protection of TMF conservation, protection of endangered species and habitats, tourism development and sustainable rural resource utilisation (Barker, 1996, 1997 and 1998). The wider objectives of the Programme were to be achieved through five main sub-programmes as follows:

- Management and co-ordination (SP-1).
- Park management (SP-2).
- Rural integrated development (SP-3).
- Environmental education (SP-4).
- Biological research and monitoring (SP-5).

The above sub-programmes were established in accordance with the EU Finance Agreement that provided funding for the Programme (Environment and Development Group, 1998).

However, the mid-term review (MTR) of the Programme concluded that it had succeeded to a limited extent in achieving its purpose of conservation through development (Environment and Development Group, 1998). In specific terms, the impact and effectiveness of SP-3 (i.e. rural integrated development) was found to be less satisfactory, largely attributed to the legacy of dubious planning and design as well as a fundamentally misconceived rural development strategy (Environment and Development Group, 1998). SP-3 was the formal attempt to put into local practice, global thinking on the support zone development programme for the Division (Chapter 4). The sub-programme had the underlying aim of reducing the pressure on the forests of the Okwangwo Division, and developing sustainable farming methods and resource extraction practices in the areas designated as the support zone of Okwangwo Division (Barker, 1998).

During the duration of the Okwangwo Programme (i.e. 1994-1998), the rural integrated development sub-programme focused on several social and economic development initiatives in the 35 villages designated as support zone (SZ) villages of the Okwangwo Division. The specific development-oriented initiatives include the development of institutional structures and systems for community participation; income generating activities; agricultural development; provision and the delivery of development infrastructures and services including roads health care, extension services (Environment and Development Group, 1998). Three of these initiatives are particularly significant to the core theme of this book. These are:

- Institutional structures for community participation:
- Alternative income generation strategies, and
- Agricultural development activities.

Institutional Structures for Community Participation
The development of institutional structures and systems for community par-

ticipation was achieved through the establishment of the three major institutions. These are:

- Village Integrated Rural Development Committees:
- Support Zone Development Association, and
- Women in Development.

The Village Integrated Rural Development Committees (VIRDCs) were established in 1995 based on the village councils and groups that were already active within the villages before the Okwangwo Programme commenced in 1994. These politically effective and responsive committees were set up to address development and conservation issues and priorities in their respective villages. Each VIRDC consisted of unpaid elected members and were organised into sub-committees to deal with particular resource management and development issues including agriculture, works, health, trade and industry. The Okwangwo Programme aimed to capture boldly, the functions and responsibilities of village councils and to set clear organisational parameters and functions that supported the aims of the Programme. The role of each sub-committee was carefully defined and documented by the Programme and was subject to signed agreements between the Programme and the members. The Okwangwo Programme had the right to withhold development inputs if agreements between the community and the Programme were broken (Environment and Development Group, 1998). This conforms to the requirement for communities to register and pledge to respect the National Park rules and regulations, as a condition for the provision of socio-economic development inputs to the villages (see Chapter 4).

The MTR of the Okwangwo Programme (Environment and Development Group, 1998) expressed concern on the sustainability of the VIRDCs. This was mainly because the Programme concentrated on setting up the institution and defining its role, but failed to adequately address the purposes of participation and the building of institutional capacity to promote long-term survival. This was further complicated by several factors including:

- Inadequate training in formal skills:
- Lack of promotion of institutional independence and initiative:
- Lack of community participation in the identification of development needs:
- Unrealistic development expectations in the villages.

As has been argued elsewhere, the Okwangwo Programme essentially adopted a top-down planning approach to conservation-with-development (Ite,

1996; Ite and Adams, 2000), and the VIRDCs served as a thin coating of community participation in this respect.

The second channel of local institutional participation in the Okwangwo Programme was through the Support Zone Development Association (SZDA), founded in December 1995 as a forum to represent the 35 support zone communities of Okwangwo Division. The original objective was to provide a link through which the Federal National Parks Service (FNPS) and the CRNP could interact with the communities around the Division. The expectation was that the SZDA would, in the long-term, become an NGO capable of developing an entity independent of any external support and representations of the SZ communities in dealings with the CRNP.

The third element of local participation in the Okwangwo Programme was Women in Development (WID). WID became the first official voice for women of the Okwangwo Division, and its survival was linked to the VIRDC as an institution. WID was successful in raising awareness of women's programmes in the SZ villages, with socio-economic development interventions including nutritional workshops and stove improvement displays. However, the MTR of the Okwangwo Programme found that WID was good in conception and in tackling real problems, but was too small in scope and scale to be efficient (Environment and Development Group, 1998). In specific terms, it lacked development tools and skills, strategic planning and target setting with the rural development component (i.e. SP-3) of the Okwangwo Programme.

Alternative Income Generating Activities
The Okwangwo Programme set up a hunter retraining scheme in 1996, to partially address the problem of alternative employment within the SZ communities. Hunters were retrained into two main trades: carpentry and tailoring. The aim was to educate people away from economic dependence on the forest; the scale of the intervention was too small to be effective due to several problems.

First, the definition of hunters by the Okwangwo Programme was too broad. It failed to differentiate between full-time and opportunistic hunters and between locals and outsiders. In addition, there was no assessment of hunter numbers, activity or identification of training potential.

Second, the overall approach in the selection and training process was paternalistic (Environment and Development Group, 1998). Trainees were subject to contractual conditions that limited their economic dependence over many years. The number of beneficiaries (a total of 15) was far too small relative to the potential number of hunters in the SZ. In addition, individuals were chosen for social reasons rather than for the scale of their actual hunting activities. As such, there was no systematic method of choosing individuals or reference to the variable economic significance of hunting within the SZ villages.

Third, no attempt was made by the Okwangwo Programme to determine demand for the particular skills and trades within the communities. In the final analysis, apprenticeships in tailoring and carpentry were selected because these trades were not well represented in the in the SZ villages, and also because placements were possible in the nearest towns, i.e. Obudu (for tailors) and Ikom (for carpenters). It is not clear the extent to which trainees could actually make a living out of their skills and how this will contribute to poverty alleviation in the SZ villages (Environment and Development Group, 1998).

Agricultural Development
The role of the Okwangwo Programme in agricultural development focused on interventions in livestock rearing and farm systems development (Chapter 4). The specific interventions in livestock rearing were aimed at increasing the population of livestock to produce an alternative protein source to bushmeat derived from the forests of the Okwangwo Division. The interventions involved providing village groups and associations of farmers with breeding stocks of pigs, sheep and goats. Healthy vaccinated animals were distributed to groups of farmers on loan, with repayment being in the form of animal offsprings returned to the Programme for distribution to other farmers. The MTR (Environment and Development Group, 1998) observed that farmers were unwilling to continue with this arrangement because of the burden of feeding animals and the need to return short-term gains to the Okwangwo Programme. All livestock recipients were members of the village elite and better-off households.

Overall, none of the livestock interventions introduced any fundamentally new technology that might lead to substantial expansion of livestock numbers in the SZ villages. According to Environment and Development Group (1998), increased livestock population was not placed in the context of a defined development opportunity in the SZ villages. In addition, the Okwangwo Programme failed to make adequate attempts towards understanding the reasons for the limited number of livestock in the SZ villages, especially since local communities were more likely to possess this knowledge (Ite, 2000b).

The intervention of the Okwangwo Programme in farm systems development involved distributing in the SZ villages various inputs and planting materials for a range of crops, including banana suckers, plantain suckers, cassava sticks and taro corms. The materials were provided free of charge, hence it was impossible to determine the extent to which there was a real demand for these products, all of which were already being cultivated by households in the Division. As such, it is probable that real demand for these planting materials could have been weak. As noted earlier, the hypothesis that there was a critical shortage of basic food in the area has been rejected (Environment and Development Group, 1998). It can be argued that the demand for plantain, banana and cassava was due to the fairly good market for the crops and farmers desire to expand production, rather than to shortfalls in household nutrition (Chapters 6 and 7).

8.6 Effectiveness and Sustainability of the Okwangwo Programme

The purpose of the EC-funded Okwangwo Programme was to protect a major portion of Nigeria's remaining primary tropical forest. This was to be achieved through the protective management and sustainable land use management (Chapter 4), while improving the living conditions of rural communities surrounding the park (Chapter 5). As shown in Chapters 5 and 6, the livelihood patterns of the SZ communities are closely linked to the forest area constituting the Okwanwgo Division of the Cross River National Park.

Against the above background, the Environment and Development Group (1998) made four major observations regarding the effectiveness and sustainability of the Okwangwo Programme. First, the Programme enhanced integrity of Okwangwo Division. Second, it improved public attitudes to the park, to conservation and the wider environment. Third, the Okwangwo Programme made inadequate material contribution to the economic and living conditions in the SZ villages; and fourth, it made insignificant contribution to sustainable land and resource use in the SZ.

These observations imply that overall, the EC-funded Okwangwo Programme succeeded in creating a strong park management framework and improved community relations and public attitudes compared to the situation during the pilot phase of conservation-with-development (Section 8.2). However, the social and economic development of the SZ villages and their livelihood systems were not sufficiently addressed by the Programme. This has significant implications for the long-term sustainability of the achievements of the Okwangwo Programme. As Environment and Development Group (1998, p.21) observed:

> The reason is that they appear to be the result, not of any real improvements in livelihoods (brought about by enhanced economic activity in the context of sustainable land and resource use), but of the programme's diverse, scattered short-term interventions. In otherwords, people have responded positively to

the mere presence of some sort of development activity in a region otherwise largely devoid of it.

This confirms the view that initial community support for the forest conservation project in Okwangwo Division was dependent on their expectations of 'real development' (Ite, 1996a) and not the cosmetic or public relations exercise provided by the rural development component (i.e. SP-3) of the Okwangwo Programme. On the other hand, it could also be argued that from the onset, it is possible that the management of the Okwangwo Programme had significant interest in biodiversity conservation more than the local socio-economic development aspects of the surrounding villages.

It is important to emphasise that the Okwangwo Programme was designed to focus on a relatively small part (i.e. the Okwangwo Division) of a single protected area (i.e. the CRNP) and not primarily concerned with institution building. Nonetheless, the Programme benefited the CRNP project as a whole, especially by allowing the FNPS to concentrate its limited and scarce resources on the Oban Division of the CRNP (see Figure 4.1). The Programme also provided an avenue for testing a model of sustainable development using the integrated conservation and development project approach, with wider implications for Nigeria and elsewhere. The Programme succeeded in generally raising the public awareness and political profile of conservation in Cross River State and Nigeria (Environment and Development Group, 1998).

In retrospect, the Okwangwo Programme faced several significant challenges in the process of combining conservation of biodiversity with rural socio-economic development in the Okwangwo Division (see Barker, 1995, 1996, 1997 and 1998). These include the deficiencies in programme preparation and design; financial and budgeting deficiencies; poor SZ community relations and the lack of appropriate human resources. Other challenges were the marginal national and local institutional support as well as the international political problems, which manifested itself in the EU and Commonwealth sanctions on Nigeria in 1995.

Against the above background, it can be argued that the sustainability of the Okwangwo Division as a protected area is dependent on several institutional factors, two of which are particularly significant. These are the ability and commitment of the FNPS to sustain the SZ village development activities and the availability of local government agencies with adequate resources to support the Okwangwo Division. On the other hand, it can be argued that the management plan for the Okwangwo Division was over-ambitious and ignored the role of state government institutions (Morakinyo, T., personal communication, April 2000). The plan attempted to create a National Park capable of undertaking all kinds of social and economic activities ranging from health care delivery to agriculture. It is believed that the Okwangwo Programme could have achieved more, if the potential role of NGOs and state government institutions were clearly acknowledged and actively incorporated during the early design phase of the park project (Morakinyo, T., personal communication, April 2000).

8.7   Conclusion

It is fairly obvious that not much was achieved by the initial implementation of global thinking on 'conservation-with-development' which was the cornerstone of the Cross River National Park project. Although the community support for the project was remarkable, this was based on local definitions, conceptions and expectations of development that were different from those of the management of the Okwangwo Division. The poor relationship between the local communities and the National Park was driven by the apparent lack of active and direct community participation in the management of the protected area. Except perhaps for the Federal National Parks Service, it is also evident that the Okwangwo Division (as an institution) was yet to gain wide institutional linkages and support within Cross River State, or elsewhere. This was mainly because opportunities for national and regional institutional support for the project were yet to be fully developed.

PART IV:
SYNTHESIS

9 Local Realities and Global Illusions

9.1 Introduction

This chapter provides a short synthesis of the main arguments presented in this book. It examines the local realities of small farmers, agriculture, forest loss and conservation in the Okwangwo Division of the Cross River National Park, Nigeria. This is followed by an analysis of the global illusions on these issues, by focusing on how the problem of tropical forest loss is framed, and the role of external agents in the conservation of local resources. The policy and research implications of the emerging issues are briefly highlighted.

9.2 Realities of Local Action

Small Farmers and Agriculture
The forest of the Okwangwo Division of the Cross River National Park in southeast Nigeria is a significant resource for households and the local economy. Households exploit the forest resource for social, economic and cultural purposes, including farming, the gathering of non-timber forest products and hunting. These activities constitute the major sources of household income. Although smallholder farming (of both cash and food crops) is the most important form of land use (Chapter 6), it is also responsible for the observed rates and patterns of forest loss (Chapter 7). Logging and timber extraction activities are of limited direct significance in the area, but under favourable conditions (e.g. prevalence of power chain saw operators) it assisted households in the rapid opening up of the forest area for the purposes of farming. The evidence presented in this book strongly suggests that several underlying variables influence households' agricultural decision-making processes, which exacerbate the rate of forest loss.

The reality is that forest loss through smallholder agriculture in the villages of the Okwangwo Division is a matter both of necessity and of choice. On the one hand, it is necessary for households to make a living from the resources at their disposal, including the forest. On the other, the choice of the best mode of utilisation of the available resource for the purposes of income generation based on households' objective functions, is grounded in complex interplay of socio-political, economic and physical environmental variables. These influence the final decision of households concerning the use of forest areas for farming. Although agricultural land use by households was established as the prime causal agent of forest loss in the study villages, it is only part of the story. Households have responded to changes in the agricultural economic variables by clearing more forest for the cultivation of banana and other food crops both for subsistence and sale. The net result has been extensive farming operations as opposed to the intensification of agricultural land use desired by forest conservation organisations and institutions.

The adoption of extensive farming methods by farmers is also affected by socio-economic and political constraints, some of which are very recent and external to the villages. The establishment of the Okwangwo Division of the Cross River National Park and the subsequent increase in the rates of forest clearance in the villages around the National Park, is a good illustration of this point. Therefore, in apportioning the blame for forest loss, it is important that the links between social, economic and political factors of deforestation are clearly identified and analysed.

Rate and Extent of Forest Loss

This book has provided new knowledge and significant insights on the patterns and rates of forest loss in the villages of the Okwangwo Division of the Cross River National Park. It is clear that there were changes in forest extent before and after the establishment of the National Park in 1991. The reality is that in the Mbe Mountains complex (a designated conservation and tourism zone of the park), approximately 38 km^2 of forest had been converted to other uses, principally agriculture, at the rate of about 1 km^2 per annum (Chapter 7). Similarly, households in the 35 support zone (SZ) villages of the Division could have cleared approximately 150 km^2 of forest, representing an annual loss rate of 6 km^2 during the period under review (i.e. 1967 to 1993). The extent of forest lost in the villages around the Division is equivalent to 15% of the total area of the Division (1,000 km^2), representing a loss rate of approximately 0.6% per annum.

The reality has been that the annual rate of forest loss (i.e. 0.6% per annum) in the Okwangwo Division is relatively modest, when compared with the annual rate of 3% reported for many parts of West Africa (Fairhead and Leach, 1998). However, it is clear that the TMF conservation initiatives for the Okwangwo Division were based on assumptions of very rapid rates of forest conversion by the local communities (Ite and Adams, 1998; Environment and Development Group, 1998). This raises two important issues (Ite, 2000b). First, the rate of forest conversion is much slower than might be feared when compared with other parts of Nigeria, particularly in the southwest. Second, there is yet no definition and agreement of what constitutes rapid rates of forest conversion for the study area.

This book questions the widely held and rather speculative notions of very high rates of forest loss in Nigeria particularly on the part of international conservation organisations. The fact that most of the villages in the Okwangwo Division and their component households have existed for more than 50 years lends further credence to the observation that forest loss in the Division has been quite a slow and gradual process. However, although forest is being lost at a slower rate than might be feared, substantial forest loss has picked up momentum in recent times due to several factors (Chapter 4). From all indications, forest covers 70% of the Division, and this consists essentially of forest in reserved areas as well as those in adjacent community-owned and managed land.

Indigenous Agro-ecological Knowledge and Agricultural Land Use
The detailed knowledge by small farmers of the vegetation types, soil characteristics, market opportunities and problems of agricultural production influence agricultural land use in sub-Saharan Africa. The practical application of this knowledge has positive and negative implications, especially for TMF loss and conservation in West Africa and Nigeria. In otherwords, the utilisation of indigenous agro-ecological knowledge by smallholder farmers in developing countries can be considered as a two-edged sword (Ite, 2000b).

On the one hand, indigenous agro-ecological knowledge has the potential to be used for sustainable forest and agricultural land use practices. The positive impact of households' utilisation of this inherent asset manifests in the cropping systems and farming processes as evident in the length of fallow periods, forest clearance techniques and the strategies for maintaining soil fertility (Chapter 6). The result has been the exploitation by households, to a relatively high level of sophistication, of the micro and macro characteristics of the local landscape. This has culminated in the relatively slow rate of TMF loss, prior to the establishment of the Okwangwo Division of the Cross River National Park. Thus, it can be argued that households' application of this knowledge has contributed significantly to sustainable local forest management, made possible by the active role of the local institutional arrangements (Chapter 5).

On the other hand, indigenous agro-ecological knowledge contributes to environmental degradation, including TMF loss. In specific terms, the utilisation of indigenous agro-ecological knowledge by households in the Okwangwo Division can be used to explain, to some extent, the observed pattern of forest clearance in the villages. It is easy for households to obtain farm plots with from the community-owned forest (Chapter 7). This has led to forest fragmentation through the selective clearance of forest patches by households based on perceptions of soil fertility and suitability for particular crops, reinforced by several forces internal and external to households and their communities. The case in point is the demise of cocoa cultivation and its more recent replacement by bananas as the main cash crop in the study area. Households are well aware of the economic returns from the cultivation of both crops. However, their economic rationality has increasingly led to banana cultivation, with significant environmental degradation, especially TMF loss. The pressing problem is how to integrate wider environmental and forest management issues into internal household economics.

It is clear from the analysis in this book that both dimensions of the potential 'value' of indigenous agro-ecological knowledge are evident in Okwangwo Division. However, on balance, it can be argued that the local institutional arrangements for forest management backed by practical utilisation of indigenous agro-ecological knowledge by households, have been responsible for the relative low rate of TMF loss in the community forests surrounding the Okwangwo Division. In otherwords, indigenous agro-ecological knowledge, as evidenced in stable, rational local cropping systems, has culminated in the relatively slow rate of TMF loss. The potential value of indigenous agro-ecological knowledge supports the current advocacy for a better understanding of the contextual causes of forest loss, especially from the small farmers' perspective (Gadgil *et al.*, 1993; Wickramasinghe, 1997).

Constraints of Agricultural Production and Development
The constraints of agricultural production and development in the support zone villages of the Okwangwo Division also constitute significant underlying factors intensifying rates and patterns of TMF loss. The problems associated with the marketing of produce and the lack of farmer support services are indirect factors contributing to forest loss. Marketing is an important component of any production system. Individual households have little or no influence on the market system, but can only respond to changes and challenges resulting from the dynamics of the system. Households will respond to unfavourable marketing terms in various ways, with the aim of counteracting the effect of low incomes arising from the unfair trading practices, but with clear implications for forest loss.

Clearly, households in forest communities do not exist in an insular economy but are linked with the wider political economy (Dove, 1983). Such linkages constitute local economic gains as well as reinforce environmental degradation. Farmers in developing countries face several problems over inputs such as agro-chemicals, which are not locally available as and when needed. Yet, these are essential for the intensification of farming operations and to some extent, a reduction in the demand for the clearing of fresh forest areas for cultivation. In effect, the problems of marketing of farm produce and the lack of extension services emerge from within the wider political economy where households find themselves.

Forest Conservation-with-Development in Practice

The forest conservation initiative in the Okwangwo Division was a step in the right direction to save the best remaining area of TMF in southeast Nigeria. However, the design and implementation of the project was based on 'conservation-with-development' ideas and principles many of which were either untried, or shown not to work well elsewhere in sub-Saharan Africa. The Okwangwo Programme attempted to stabilise agricultural land use around the Okwangwo Division of the Cross River National Park by providing compensation and incentives to the local communities (Chapters 4 and 8).

From a conservation perspective, the Programme succeeded in creating a strong park management framework for the purposes of protecting the forests of the Division. From the socio-economic development perspective, the short to medium-term results were not impressive. Initially, few households appreciated the benefits the park offered, while the majority became dissatisfied with the immediate (and most likely long-term) disadvantages of the National Park project. These included the perceived reduction in agricultural land area as well as losses of traditional hunting areas, fishing rights, gathering areas and access to economic trees. Communities initially supported the forest conservation project because they expected benefits that were significantly different from those planned by the management of the National Park and Okwangwo Programme. In many respects, some households and communities regarded the entire National Park project as an aid agency and a development institution.

The reality was that the degree of household and community support for external forest conservation initiatives was directly related to the perception of the potential benefits of such projects relative to current traditional uses of land. Most TMF conservation projects in developing countries are based on the philosophy of 'conservation-with-development' (Chapter 1). It many cases, local communities always constitute peripheral elements during the project formulation or planning phase. Their support tends to be sought during the implementation phase of such projects. As Caldecott (1996, pp.57) admitted for the Cross River National Park project in Nigeria, there was:

> ... a lack of specific and relevant experience among all the groups and most of the individuals involved in planning the project. The group with most relevant experience in fact probably comprised the local people in the project area,

who had been many times exposed to government plans and projects affecting the forests and lands which they regard as their own. This was not fully reflected in either the consultation or the planning process that defined the project.

There is no doubt that the Okwangwo Programme had the support of international environmental organisations and development agencies. However, internal support was not fully gained and consolidated at the national and regional levels within Nigeria. It has been suggested that the Programme ignored state development institutions and local NGOs who were (or could be have been) potential partners in the support zone development activities. This suggests that although the forest conservation project had been an externally conceived idea at the planning phase, it largely remained so during the implementation stage. In many respects, the EC-funded Okwanwgo Programme (Chapter 8) was a top-bottom approach to conservation-with-development with a thin coating of 'local participation'. Without adequate local support, the translation of global thinking on forest conservation into concrete local action in southeast Nigeria is likely to remain a challenge for conservation organisations and practitioners with significant interest in the forest of the Okwangwo Division.

This book therefore challenges very optimistic descriptions of successful 'conservation-with-development' projects in developing countries (Chapter 1), especially in tropical moist forest areas. The development and management of National Parks based on the principles of conservation-with-development are more likely to succeed in conserving TMF environments, if they are able to meet the legitimate socio-economic development aspirations of the people living in and around such areas.

9.3   Illusions in Global Thinking

There is no doubt that the human ecological and political economic perspectives dominate global thinking on the causes of tropical forest loss (Chapter 2). Although there is some degree of correlation between demographic factors and rates of tropical forest loss in developing countries, it is pertinent to acknowledge that this relationship is also influenced by several factors. These include political, social and economic factors operating within and outside the immediate environment of smallholder farming households in developing countries.

The choice of extensive agricultural land use systems by households in developing countries is largely a function of households' perception of the availability of abundant forest resource and the economic returns from the exploitation of forest resources. The creation of National Parks and the incorporation of previously community-owned and managed forest areas into the National Park can reduce the forest extent of the surrounding villages. This can result in increased rates of forest clearance by households, partly in anticipation of the response by conservation project management that would be detrimental to the interests of both households and local communities. In otherwords, forest loss can be driven by external intervention in the traditional management of forest resources by local communities. This has been the case in the Okwangwo Division of the Cross River National Park. As Ite and Adams (1998) observed, conservation activity, stimulated by a generalised and exaggerated picture of the causes and rates of local forest loss, might actually be contributing to the problem it was trying to solve.

It is clear that several underlying factors influence household decisions in clearing forest areas for farming. The tropical forest provides an important resource base for households, with farming constituting a significant income opportunity. At the household level, household farming decisions are influenced by factors such as the distribution and size of holdings, land tenure, availability of labour and capital equipment. The emerging cropping systems in tropical forests are also determined to a large extent by soil fertility levels as well as the constraints of agricultural production. Thus any observed patterns and rates of forest loss are enmeshed in a complex web of environmental, social, political and economic variables. As such, human ecology and political economy interact in explaining forest loss at the local level.

The empirical evidence presented in this book challenges the analysis of the causes of forest loss which attribute the problem solely to either demographic or political economic factors. Arguably, at the national level such an analysis might provide insights to aggregate processes of deforestation. However, at the local level (village and household) such as reflected in this book, it is highly unlikely that the human ecological arguments or the political economic perspective would hold much water as the sole explanation of changes in forest extent, and subsequent forest loss.

It is evident that global policy responses to the problem of forest loss in developing countries have been based on the dominant perspectives on the causes of the problem discussed above (see also Chapter 2). This has resulted in the design and implementation of 'top-down' forest conservation projects, often with significant financial support from the international community. Realistic international policies for controlling tropical forest loss can only be derived through the appropriate framing of the problem at the local level. This reality is yet to be developed and implemented.

9.4 Conclusion

From a theoretical perspective, this book has contributed to the debates on the explanation of the dynamics and patterns of TMF loss. This requires the study of a representative number of areas of forest loss since deforestation occurs in a small number of high-intensity fronts (Sierra, 2000). The materials presented in this book also underlines the need for caution in the interpretation, use and analysis of data on the causes and consequences of tropical deforestation. As Ochoa-Gaona and Gonzalez-Espinosa (2000) observed, in some cases, the use and analyses of data on tropical forest loss frequently neglect many aspects and scales of the processes of loss within a given region.

This book challenges the conventional view of the 'irrational behaviour' of small farmers and their role in forest loss. It is very fashionable for government departments, international development agencies and conservation organisations, to blame the 'uneducated farmer' in developing countries for TMF loss rather than an aggressive farming industry, a faltering government institution or elite opportunism. Yet, the relationship between small farmers and the forest is relatively less understood. It is evident that current explanations of the role of small farmers in TMF loss have failed to recognise that smallholder agriculture, agroforestry and pastoral systems can be ecologically sustainable. Most smallholders are rational in decision making and possess adaptive behaviour and skills based on their understanding of the local environment. In addition, arguments about the destructive influence of small farmers often underestimate the role of external factors such as government policies, class position and land tenure, in the internal (household) organisation of resource management. Yet these issues play significant roles in influencing household decision-making concerning land use and the management of the forest environment.

The book contends that present global generalisations of the relationship between population growth and forest loss have limited value, especially for understanding local (household) processes of forest loss. In some places it would be readily obvious that global demand for timber drives forest loss, while in other geographical locations, population growth could be the major factor. Clearly, TMF loss has no simple single cause and the role of smallholder farmers varies considerably from place to place through time.

This book has gone beyond the conventional generalised treatment of the role of farmers in TMF loss, which essentially fails to consider the internal dynamics of the individual households within the agrarian economy. It offers a new perspective on the role of small farmers in TMF loss by suggesting that decision-making models of agrarian change are ideal for understanding and explaining the problem of forest loss at the local (household) level. The causes, rate and location of farmer encroachment on the forest cannot be understood without considering decision-making in farming households against the wider context of the economy and certainly not in terms of a simple neo-Malthusian and population growth model. There is therefore an increasing need for more contextual analysis of the causes of TMF loss from the households' perspective. The political ecology approach (Chapter 2) has the potential to serve as a very useful and appropriate analytical framework.

This book does not lay any claim to the wholesale adoption of the political ecology approach in its analysis of the problem of tropical forest loss and the incorporations of small farmers in conservation policy and planning. Rather, it highlights and recognises the superiority of the approach for explaining the causes of land degradation, including forest loss. Only one dimension (i.e. the household) of the political ecology approach has been explored in this book. It shows the dynamic economic situation for farmers, and stresses the need to study them at the level of the individual household. Although the larger-scale political economy of production, including inter-regional and international trade in forest and agricultural products, macro-economic policy, debt are very important especially in Nigeria, they are not explored in this book, and must make the subject of further research and publication. Nonetheless, the approach which was adopted in the research for this book (Ite, 1995), does fit broadly within the political ecology perspective. The political ecology approach is also relevant for articulating the role of small farmers in forest conservation. If global conservation agents misunderstand the dynamics of the encroachment on the forest by local small farmers, responses to such encroachment will be ineffective. Furthermore, if such responses are top-down or badly organised, these could exacerbate the problem of forest loss.

From an empirical perspective, this book has provided a detailed account of the tensions between agriculture, small farmers and tropical forest conservation in southeast Nigeria. There are significant strains in the relationship between conservation and development in tropical forest areas across the world. This is evident in the number of 'conservation-with-development' projects which fail to deliver adequate results simultaneously on both conservation and development aspects of tropical forest management (Oates, 1999). The small farmers in southeast Nigeria view the TMF environment as the source of livelihood for the present and future generations. These farmers are indeed conservationists, and do not cause permanent degradation of soils deliberately, except in extreme circumstances, or accidentally. To succeed, forest conservation policies for developing countries must incorporate the knowledge of small farmers, especially during the design phase of 'conservation-with-development' projects. Such knowledge can only be garnered through micro-level case studies and long-term research, with individual farming units (households) as the focus of the analysis.

Future forest conservation initiatives in developing countries are more likely to achieve their goals if their managers are willing and prepared to learn from the local resource users. The existing and very uncomfortable gap between global thinking (theory) and local action (practice) on tropical forest conservation can be easily reduced (if not completely eliminated) through genuine local participation by all stakeholders in forest management initiatives.

Bibliography

Abasiattai, M.B. (1987), 'History of Cross River State', in M.B. Abasiattai (ed.), *Akwa Ibom and Cross River States: The Land, the People and their Culture*, Wusen Press, Calabar, Nigeria, pp.47-69.

Abumere, S.I. (1983), 'Traditional agricultural systems and staple food production', in J.S.Oguntoyinbo, O.O. Areola and M. Filani (eds), *A Geography of Nigerian Development*, 2nd Edition, Heinemann Educational Books (Nigeria) Limited, Ibadan, pp.244-261.

Adams, J.S. and McShane, T.O. (1992), *The Myth of Wild Africa: Conservation without Illusion*, W.W. Norton, London.

Adams, W. M. (1990), *Green Development: Environment and Sustainability in the Third World*, Routledge, London.

Adeola, M.O. (1992), 'Importance of wild animals and their parts in the culture, religious and traditional medicine of Nigeria', *Environmental Conservation*, vol.19, pp.125-134.

Allen, J.C. and Barnes, D.F. (1985), 'The causes of deforestation in developing countries', *Annals of the Association of American Geographers*, vol.75, pp.163-184.

Anadu, P.A. (1987), 'Progress in the conservation of Nigeria's wildlife', *Biological Conservation*, vol. 41, pp.237-251.

Anderson, D. and Grove, R. (eds) (1987), *Conservation in Africa: People, Policies and Practice*, Cambridge University Press, Cambridge.

Angelsen, A. (1999), 'Agricultural expansion and deforestation: modelling the impact of population, market forces and property rights', *Journal of Development Economics*, vol.58, pp. 185-218.

Angelsen, A. and Kaimowitz, K. (1999), 'Rethinking the causes of deforestation: lessons from economic models', *The World Bank Research Observer*, vol.14, no.1, pp73-98.

Anthonio, Q.B.O. (1990), *Observations on Agricultural Economics in the Support Zone*, WWF-UK, Goldalming.

Armstrong, G., Fawcett, T., White, J. and Okali, D. (1990) Cross River Forestry Project, Nigeria: Report on a Project Preparation Visit. 17 January - 8 February 1990. Unpublished, ODA.

Ashton-Jones, N. (1992), *Project Manger's Report: ODA Project (WWF No. 3916) Development of the Northern Sector of the Cross River National Park*, WWF-CRNPP, Calabar. Unpublished Report to ODA, London.

Atte, O.D. (1994), *Lands and Forests of Cross River State: A Participatory Appraisal of Rural Peoples Perceptions and Preferences*, Cross River State Forestry Project (ODA Assisted) Working Paper no. 9, Forestry Department Headquarters, Calabar.

Barker, J. (1995), *Cross River National Park (Okwangwo Division) Annual Report, 1 July 1994 – 30 June 1995*, WWF No. NGO003, DFID-JFS No.321, WWF-UK, Godalming.

Barker, J. (1996), *Cross River National Park (Okwangwo Division) Annual Report, 1 July 1995 – 30 June 1996*, WWF No. NGO003, DFID-JFS No.321, WWF-UK, Godalming.

Barker, J. (1997), *Cross River National Park (Okwangwo Division) Annual Report, 1 July 1996 – 30 June 1997*, WWF No. NGO003, DFID-JFS No.321, WWF-UK, Godalming.

Barker, J. (1998), *Cross River National Park (Okwangwo Division) Annual Report, 1 July 1997 – 30 June 1998*, WWF No. NGO003, DFID-JFS No.321, WWF-UK, Godalming.

Barrett, C.S. and Arcese, D.P. (1995), 'Are integrated conservation-development projects (ICDPs) sustainable? On the conservation of large mammals in Sub-Saharan Africa', *World Development*, vol.23, pp.1073-1084.

Basset, T.J. (1988), 'The political ecology of peasant-herder conflicts in northern Ivory Coast', *Annals of the Association of American Geographers*, vol.78, pp.453-472.

Bayliss-Smith, T.P. (1982), *The Ecology of Agricultural Systems*, Cambridge University Press, Cambridge.

Bilsborrow, R. and Geores, M. (1994), 'Population, land use and the environment in developing countries: what can we learn from cross-national study?', in K. Brown and D. Pearce (eds) (1994), *The Causes of Tropical Deforestation*, University of London Press, pp.106-133.

Blaikie, P. (1989), 'Explanation and policy in land degradation and rehabilitation for developing countries', *Land Degradation and Rehabilitation*, vol.1, pp. 23-37.

Blaikie, P.M. (1985), *The Political Economy of Soil Erosion in Developing Countries*, Longman, London.

Blaikie, P.M. and Brookfield, H. (1987), *Land Degradation and Society*, Methuen, London.

Blench, R., Marriage, Z. and Morakinyo, T. (1999), *Background Paper: Environment Nigeria*, Overseas Development Institute, London.

Boserup, E. (1965), *Conditions of Agricultural Growth*, George Allen and Unwin Limited, London.

Bourke, G. (1987), 'Forests in the Ivory Coast face extinction', *New Scientist*, 11 June, p.22.

Brandt Commission (1983), *Our Common Crisis*, Pan Books, London.

Brechin, S.R., West, P.C., Harmon, D. and Kutay, K. (1991), 'Resident peoples and protected areas: a framework for inquiry', in P.C. West and S.R. Brechin (eds), *Resident Peoples and National Parks: Social Dilemmas and Strategies in International Conservation*, The University of Arizona Press, Tucson, pp.5-28.

Brown, J.L (1991), *Building Community Support for Protected Areas: The Case of Tortuguero National Park, Costa Rica*, Unpublished MA Thesis, Clark University, USA.

Brown, J.L. and Brenes, H. (1992), 'A Strategy to Build Community Support for Tortuguero National Park, Costa Rica: Some Recommendations', Paper presented at the IV World Congress on National Parks and Protected Areas, Caracas, Venezuela, February.

Brown, K. (1998), 'The political ecology of biodiversity, conservation and development in Nepal's Terai: confused meanings, means and ends', *Ecological Econom-ics*, vol. 24, pp. 73-87.

Brown, K. and Pearce, D. (eds) (1994), *The Causes of Tropical Deforestation*, University of London Press.

Bryant, R.L. (1992), 'Political Ecology: An Emerging Research Agenda in Third World Studies', *Political Geography*, vol.11, pp. 12-36.

Byron, N. and Arnold, M. (1999), 'What Futures for the People of the Tropical Forests?', *World Development*, vol.27, pp. 789-805.

Caldecott, J. (1996), *Designing Conservation Projects*, Cambridge University Press, Cambridge.

Caldecott, J. and Morakinyo, A.B. (1996), 'Nigeria', in E. Lutz and J. Caldecott (eds), *Decentralization and Biodiversity Conservation*, The World Bank, Washington D.C., pp. 78-90.

Caldecott, J. O., Bennet, J.G. and Ruitenbeek, H.J. (1989), *Cross River National Park (Oban Division): Plan for Developing the Park and its Support Zone*, WWF-UK/ODNRI, Godalming.

Caldecott, J. O., Oates, J.F. and Ruitenbeek, H.J. (1990a), *Cross River National Park (Okwangwo Division): Plan for Developing the Park and its Support Zone*, WWF-UK, Godalming.

Caldecott, J. O., Oates, J.F., Gadsby, E.L. and Edet, C.A. (1990b), *Gorilla-based and other Conservation Development Potential*, WWF-UK, Godalming.

Cartwright, J. (1991), 'Is there hope for conservation in Africa?', *The Journal of Modern African Studies*, vol.29, pp. 355-371.

Clay, J.W. (1988), *Indigenous Peoples and Tropical Forests: models of land use and management for Latin America*, Cultural Survival Inc., Cambridge, Massachussetts.

Coomes, O.T., Grimard, F. and Burt, G.J. (2000), 'Tropical forests and shifting cultivation: Secondary forest fallow dynamics among traditional farmers of the Peruvian Amazon', *Ecological Economics*, vol. 32, pp.109-124.

Cross River Agricultural Development Project (CRADP) (1986), *On-Farm Adaptive Research: Diagnostic Survey of 60,000 Farm Families in Cross River State*, Unpublished CRADP Studies, Calabar, Nigeria.

Cross River Forestry Department (CRFD) (1988), *Annual Report*, Forestry Development Headquarters, Calabar, Nigeria.

Cross River Forestry Department (CRFD) (1989), *Annual Report*, Forestry Development Headquarters, Calabar.

Cross River Forestry Department (CRFD) (1990), *Annual Report*, Forestry Development Headquarters, Calabar.

Cross River Forestry Department (CRFD) (1991), *Annual Report*, Forestry Development Headquarters, Calabar.

Cross River Forestry Department (CRFD) (1992), *Annual Report*, Forestry Development Headquarters, Calabar.

Cross River Forestry Department (CRFD) (1994), 'Papers for the Second Meeting of the Steering Council for the Cross River State Forestry Sub-sector Strategic Plan', Metropolitan Hotel, Calabar, 20th July, Cross River Forestry Development Department.

Cross River Forestry Department (CRFD) and Overseas Development Administration (ODA) (1994), *Technical Report: An Overview of a Planning Process for Sustainable Management of the Forest of Cross River State, Nigeria*, Cross River Forestry Department/Overseas Development Administration, London.

Dasmann, R. F. (1984), 'The relationship between protected areas and indigenous people', in J.A.McNeely and K.R. Miller (eds), *National Parks, Conservation and Development: The Role of Protected Areas in Sustaining Society*, IUCN/ Smithsonian Institution Press, pp. 667-671.

Davies, G. and Richards, P. (1991), 'Rain Forest in Mende Life: Resources and Subsistence Strategies in Rural Communities and the Gola North Reserve (Sierra Leone)', Report to ESCOR, UK Overseas Development Administration, London.

Dixon, J. A. and P. B. Sherman (1990), *Economics of Protected Area: A New Look at Benefits and Cost*, Earthscan, London.

Dove, M.R. (1983), 'Theories of swidden agriculture and the political economy of ignorance', *Agroforestry Systems*, vol.1, pp. 85-99.

Dunn, R., Otu, D.O., Wong, J.L.G. (1994), *Report of the Reconnaissance Inventory of the High Forest and swamp Forest Areas in Cross River State, Nigeria*, CRSFP (ODA Assisted), Forestry Department Headquarters, Calabar.

Ebin, C. (1992), 'Brief on the Cross River National Park', WWF-CRNPP, Calabar, 4pp.

Ekong, E.E. (1987), 'Traditional political system in Cross River State', in M.B. Abasiattai (ed), *Akwa Ibom and Cross River States: The Land, the People and their Culture*, Wusen Press, Calabar, Nigeria, pp. 89-103.

Ekpo, A.H. (1993), 'An overview of economic growth and development', in I.A. Adalemo and J.M. Baba (eds) *Nigeria: Giant in the Tropics, Vol. 1, A Compendium*, Gabumo Publishing, Lagos, pp.111-116.

Environment and Development Group (EDG) (1998), *The Federal Republic of Nigeria: Mid Term Review of the Okwangwo Programme*, Draft Report, May 1998, EDG. Oxford.

Essien, O.E. (1987), 'Cross River State languages: problems and prospects', in M.B. Abasiattai (ed), *Akwa Ibom and Cross River States: The Land, the People and their Culture*, Wusen Press, Calabar, Nigeria, pp. 27-45.

Etta, C.N.H. (1987), 'Message from the Programme Manager, Cross River Agricultural Development Project', *CRADP Newsletter*, Maiden Issue, Oct-Dec, 1987, CRADP, Calabar.

Fairhead, J. and Leach, M. (1994), 'Contested forests: modern conservation and historical land use in Guinea's Ziama Reserve', *African Affairs*, vol.93, pp. 481-512.

Fairhead, J. and Leach, M. (1998), *Reframing Deforestation: Global Analysis and Local Realities: Studies in West Africa*, Routledge, London.

Federal Environmental Protection Agency (FEPA) (1992), *Country Study Report for Nigeria on Costs, Benefits and Unmet Needs of Biological Diversity Conservation*, National Biodiversity Unit, The Presidency, Abuja, Nigeria.

Federal Republic of Nigeria (1991a), 'Inauguration of the Natural Resources Conservation Council: An Address by the President, Commander-in-Chief of the Armed Forces, General Ibrahim Badamasi Babangida, CFR, Fss, mni', 1st February 1991, Lagos, Nigeria.

Federal Republic of Nigeria (1991b), *Decree 36 - National Parks Decree 1991, Offi-*

cial Gazette No.4 Vol. 78, Federal Government Printer, Lagos.

Fletcher, S.A. (1990), 'Parks, protected areas and local populations: new international issues and imperatives', *Landscape and Urban Planning*, vol.19, pp. 197-201.

Forrest, T. (1993), *Politics and Economic Development in Nigeria*, Westview Press, Boulder.

Fujisaka, S. (1989), 'The need to build upon farmer practice and knowledge: reminders from selected upland conservation projects and policies', *Agroforestry Systems*, vol.9, pp. 141-153.

Gadgil, M., Berkes, F., Folke, C. (1993), 'Indigenous knowledge for biodiversity conservation', *Ambio*, vol.22, pp.151-156.

Galletti, R., Baldwin, K.D.S. and Dina, I.O. (1956), *Nigerian Cocoa Farmers: An Economic Survey of Yoruba Cocoa Farming Families*, Oxford University Press.

Gordon, R.J. (1985), 'Conserving bushmen to extinction in southern Africa', *An End to Laughter? Tribal Peoples and Economic Development*, Review No. 44, Survival International, London.

Gourou, P. (1980), *The Tropical World*, Longman, London.

Gradwohl, J. and Greenberg, R. (1988), *Saving the Tropical Forests*, Earthscan, London.

Grainger, A. (1993), 'Rates of deforestation in the humid tropics: estimates and measurement', *The Geographical Journal*, vol.159, pp. 33-44.

Grove, R. (1987), 'Early themes in African conservation: the Cape in the nineteenth century', in D. Anderson and R. Grove (eds), *Conservation in Africa: People, Policies and Practice*, Cambridge University Press, pp. 21-40.

Grove, R. (1990), 'The origins of environmentalism', *Nature, vol.*345, no.6270, pp. 11-14.

Guppy, N. (1983), 'The case for an Organisation for Timber Exporting Countries (OTEC)', in S.L. Sutton, T.C. Whitmore and A.C. Chadwick (eds), *Tropical Rain Forest: Ecology and Management*, Blackwell Scientific Publishers, pp. 459-463.

Hales, D. (1989), 'Changing concepts of National Parks', in D. Western and M. Pearl (eds), *Conservation for the Twenty-first Century*, Oxford University Press, pp.139-144.

Hamilton, C.S. (1984), *Deforestation in Uganda*, Oxford University Press.

Hannah, L. (1992), *African People, African Parks: An Evaluation of Development Initiatives as a Means of Improving Protected Area Conservation in Africa*, USAID/Biodiversity Support Program and Conservation International, USA.

Harcourt, A.H., Pennington, H. and Weber, A.W. (1986), 'Public attitudes to wildlife and conservation in the Third World', *Oryx*, vol.20, pp. 152-154.

Harcourt, A.H., Stewart, K.J. and Inahoro, I.M. (1989), 'Gorilla quest in Nigeria, *Oryx*, vol.23, pp. 7-13.

Harriss, J. (1982), 'General introduction', in J. Harriss (ed) *Rural Development: Theories of Peasant Economy and Agrarian Change*, Routlegde, pp.15-34.

Hecht, S. (1985), 'Environment, development and politics: capital accumulation and the livestock sector in Eastern Amazonia', *World Development*, vol.13, pp. 663-684.

Helleiner, G.K. (1966), *Peasant Agriculture, Government and Economic Growth in*

Nigeria, The Economic Growth Center, Yale University.

Holland, M.D., Allen, R.K.G., Barton, D and Murphy, S.T. (1989) *Land Evaluation and Agricultural Recommendations*, WWF-UK/ODNRI, Godalming.

Holmgren, P., Masakha, E.J. and Sjoholm, H. (1994), 'Not all African land is being degraded: a recent survey of trees on farms in Kenya reveals rapidly increasing forest resources', *Ambio*, vol.23, pp. 390-395.

Hough, J. L. (1988), 'Obstacles to effective management of conflicts between national parks and surrounding human communities in developing countries', *Environmental Conservation*, vol.15, pp.129-136.

Hurst, F. and Thompson, H. (1994), 'A Review of the Okwangwo Division of the Cross River National Park' Final Report, Unpublished, WWF-UK, Godalming.

International Union for the Conservation of Nature (IUCN) (1980), *The World Conservation Strategy*, IUCN/WWF/ UNEP, Gland, Switzerland.

International Union for the Conservation of Nature (IUCN) (1985), *UN List of National Parks and Protected Areas*, IUCN, Gland, Switzerland.

International Union for the Conservation of Nature (IUCN) (1986), *Review of the Protected Area System in the Afrotropical Realm*, IUCN, Gland, Switzerland.

International Union for the Conservation of Nature (IUCN) (1987), *Action Strategy for Protected Areas in the Afrotropical Realm*, IUCN, Gland, Switzerland.

International Union for the Conservation of Nature (IUCN) (1994a), *Guidelines for Protected Area Management Categories*, IUCN, Gland, Switzerland.

International Union for the Conservation of Nature (IUCN) (1994b), *1993 UN List of National Parks and Protected Areas*, IUCN, Gland, Switzerland.

Ite, U.E. (1995), 'Agriculture and Tropical Forest Conservation in Southeast Nigeria', Unpublished PhD Dissertation, University of Cambridge.

Ite, U.E. (1996a), 'Community Perceptions of the Cross River National Park, Nigeria', *Environmental Conservation*, vol.23, pp. 351-357.

Ite, U.E. (1996b), 'Small farmers and tropical forest loss: towards an alternative theoretical perspective', Paper presented at the Conference on Tropical Forest Research: Current Directions and Perspectives, RGS/IBG, London: October.

Ite, U.E. (1997), 'Small farmers and forest loss in Cross River National Park, Nigeria', *The Geographical Journal*, vol.163, pp. 47-56.

Ite, U.E. (1998), 'New wine in an old skin: the reality of tropical moist forest conservation in Nigeria', *Land Use Policy*, vol.15, pp. 135-147.

Ite, U.E. (2000a), 'Evolution and Sustainability of Intermediate Systems: Tree Integration in Homestead Farms in Southeast Nigeria', Paper presented at the Workshop on Cultivating (in) Tropical Forests: the evolution and sustainability of intermediate systems between extractivism and plantations, Lofoten, Norway, 28June-1July. CIFOR/FORESSASIA/ETFRN.

Ite, U.E. (2000b), 'Indigenous agro-ecological knowledge, tropical forest loss and conservation in southeast Nigeria', Paper presented at the Workshop on Learning from Resource Users – A Paradigm Shift in Tropical Forestry? ANN-ETFRN Vienna, Austria, 28-29 April.

Ite, U.E and Adams, W.M. (1998), 'Forest conversion, conservation and forestry in Cross River State, Nigeria', *Applied Geography*, vol.18, pp. 301-314.

Ite, U.E. and Adams, W.M. (2000), 'Expectations, impacts and attitudes: conservation and development in Cross River National Park, Nigeria', *Journal of International Development*, vol.12: pp. 325-342.

Jagannathan, N.V. and Agunbiade, A.O. (1990), *Poverty-Environment Linkages in Nigeria: Issues for Research*, Policy and Research Division Working Paper No. 1990-7, Environment Department, The World Bank.

Jain, S. (1984), 'Standing up for trees: women's role in the Chipko movement', *Unasylva*, vol.36, pp. 12-20.

Kalipeni, E. and Oppong, J. (1998), 'The refugee crisis in Africa and implications for health and disease: a political ecology approach', *Social Science and Medicine*, vol.46, pp.1637-1653.

Kummer, D. and Sham, C.H. (1994), 'The causes of tropical deforestation: a quantitative analysis and case study for the Philipines', in K. Brown and D. Pearce (eds) (1994), *The Causes of Tropical Deforestation*, University of London Press, pp. 146-158.

Kramer, R.A., Sharma, N., Shyamsundar, P. and Munasinghe, M. (1994), *Costs and Compensation Issues in Protecting Tropical Rainforests: Case Study of Madagascar*, Environment Working Paper No. 62, Africa Technical Department, Environment Department, The World Bank.

Lado, C. (1992), 'Problems of wildlife management and land use in Kenya', *Land Use Policy*, vol.9, pp. 169-184.

Lagemann, J. (1977), *Traditional African Farming Systems In Eastern Nigeria: An Analysis of Reaction to Increasing Population Pressure*, Weltforum Verlag, Munchein.

Lal, R. and Okigbo, B. (1990), *Assessment of Soil Degradation in the Southern States of Nigeria*, Environment Working Paper No. 39, Environment Department, The World Bank.

Lanly, J.P. (1982), *Tropical Forest Resources*, FAO Forestry Paper No. 30, FAO, Rome.

Larson, P.S., Freudenberger, M. and Wyckoff-Baird, B. (1998), *WWF Integrated Conservation and Development Projects: Ten Lessons from the Field 1985-1996*, World Wildlife Fund, Washington, D.C.

Laurance, W.F. (1999), 'Reflections on the Tropical Deforestation Crisis', *Biological Conservation*, vol.91, pp.109-117.

Leach, M. (1994), *Rainforest Relations: Gender and Resource Use among the Mende of Gola, Sierra Leone*, Edinburgh University Press.

Leader-Williams, N., Harrison, J. and Green, M.J.B. (1990), 'Designing protected areas to conserve natural resources', *Scientific Progress, Oxford*, vol.74, pp. 189-204.

Lewis, D., Kaweche, G.B. and Mwenya, A. (1990), 'Wildlife conservation areas outside protected areas - lessons from and experiment in Zambia', *Conservation Biology*, vol.4, pp. 171-180.

Lindsay, W.K. (1987), 'Integrating, park and pastoralists: some lessons from Amboseli', in D. Anderson and R. Grove (eds) *Conservation in Africa: People, Policies and Practice*, Cambridge University Press, pp. 149-167.

Livingstone, I. (1979), 'On the concept of integrated rural development planning in less developed countries', *Journal of Agricultural Economics*, vol.30, pp. 49-53.

Lowe, R. (1992), 'Nigeria', in Sayer *et al.* (eds), *The Conservation Atlas of Tropical Forest: Africa*, IUCN/WCMC, pp.230-239.

Lowe, R.G. (1984), 'Forestry and forest conservation in Nigeria', *Commonwealth Forestry Review*, vol.63, pp. 129-136.

Lusigi, W.J. (1981), 'New approaches to wildlife conservation in Kenya', *Ambio*, vol.10, pp. 2-3.

MacDonald, M.N. (1994), 'An Assessment of Health Needs and Conditions in Support Zone Villages, Okwangwo Division, Cross River National Park', Report to WWF-CRNPP, Calabar.

MacKenzie, J.M. (1987), 'Chivalry, social Darwinism and ritualised killing: the hunting ethos in Central Africa up to 1914', in D. Anderson and R. Grove (eds) *Conservation in Africa: People, Policies and Practice*, Cambridge University Press, pp. 41-62.

MacKinnon, J., MacKinnon, K., Graham, C. and Thorsell, J. (1986), *Managing Protected Areas in the Tropics*, IUCN/UNEP.

Marshall, P. (1993a), 'Project Manager's Report - July, August and September 1993', Unpublished, WWF-CRNPP, Calabar.

Marshall, P. (1993b), 'Cross River National Park (Okwangwo Division): Development of the Northern Sector of the CRNPP. Report for the period 1st April 1992 to 31st March 1993', Unpublished, WWF-CRNPP, Calabar.

Marshall, P. (1994a), 'Project Manager's Report - October, November and December 1993', Unpublished, WWF-CRNPP, Calabar.

Marshall, P. (1994b), 'Project Manager's Report - January, February and March, 1994', Unpublished, WWF-CRNPP, Calabar.

Martin, M. (1991), *The Rainforests of West Africa: Ecology - threats - conservation*, Birkhauser Verlag, Basel, Boston, Berlin.

Mascarenhas, A. (1983), 'Ngorongoro: a challenge to conservation and development', *Ambio*, vol.12 (3-4), pp. 146-152.

Matowanyinka, J.Z.Z. (1992), 'Linking Human and Biological Aspects with respect to Protected Areas: Some Observations on Indigenous Systems in Africa', Paper presented at the IV World Congress on National Parks and Protected Areas, Caracas, Venezuela, February.

McNeely, J.A. (1988), *Economics and Biological Diversity: Developing and Using Economic Incentives to Conserve Biological Resources*, IUCN, Gland, Switzerland.

McNeely, J.A. (1989), 'Protected areas and human ecology: how National Parks contribute to sustaining societies in the twenty-first century', in D. Western and M. Pearl (eds), *Conservation for the Twenty-first Century*, Oxford University Press, pp.150-157.

McNeely, J. A. and Miller, K.R. (1983), 'IUCN, National Parks and Protected Areas: priorities for action', *Environmental Conservation*, vol.10, pp. 13-21.

McShane, T.O. (1990), 'Wildlands and human needs: resource use in an African protected area', *Landscape and Urban Planning*, vol.19, pp. 145-148.

Mkanda, F.X. and Munthali, S.M. (1994), 'Public attitudes and needs around Kasungu National Park, Malawi', *Biodiversity and Conservation*, vol.3, pp. 29-44.

Morakinyo, T. (1992), *The History of Deforestation in Nigeria, 1400-1990s*, CRSFP/

CRNPP, Calabar.

Myers, N. (1980), *Conversion of Tropical Moist Forests*, National Academy of Sciences, Washington D.C.

Myers, N. (1984), *The Primary Source: Tropical Forest and Our Future*, W.W. Norton, New York.

NARESCON (1992), *Natural Resources Conservation Action Plan: Final Report Vol. 1.*, National Advisory Committee on Conservation of Renewable Resources, The Presidency, Abuja, Nigeria.

Neumann, R.P. (1992), 'Political ecology of wildlife conservation in the Mt. Meru area of northeast Tanzania', *Land Degradation and Rehabilitation*, vol.3, pp. 85-98.

Newmark, W.D. (1993), 'The role and design of wildlife corridors with examples from Tanzania', *Ambio*, vol.22, pp. 500-504.

Newmark, W.D., Leonard, N.L., Sariko, H.I., and Gamassa, D.M. (1993), 'Conservation attitudes of local people living adjacent to five protected areas in Tanzania', *Biological Conservation*, vol.63, pp. 177-183.

Nextoux, F. and Dudley, N. (1987), *A Hard Wood Story: An Investigation into the European Influence on Tropical Forest Loss*, Friends of the Earth Trust/Earth Resources Ltd, UK.

Nigeria Environmental Study/Action Team (NEST) (1991), *Nigeria's Threatened Environment: A National Profile*, NEST, Ibadan.

Nwafor, J.C. (1982), 'Agricultural zones', in K.M. Barbour, J.S. Oguntoyinbo, J.O.C. Onyemelukwe and J.C. Nwafor (eds), *Nigeria in Maps*, Hodder and Stoughton, London.

Nye, P.H. and Greenland, D.J. (1961), *The Soil Under Shifting Cultivation*, Commonwealth Agricultural Bureaux Technical Communication No. 51, England.

Oates, J.F. (1995), 'The dangers of conservation by rural development - a case study from the forests of Nigeria', *Oryx*, vol.29, pp.115-122.

Oates, J.F. (1999), *Myth and Reality in the Rain Forest: How Conservation Strategies are Failing in West Africa*, University of California Press, Berkeley.

Oates, J.F., White, D., Gadsby, E.L. and Bisong, P.O. (1990), *Conservation of Gorillas and Other Species*, WWF-UK, Godalming.

Ochoa-Gaona, S. and Gonzalez-Espinosa, M. (2000), 'Land use and deforestation in the highlands of Chiapas, Mexico', *Applied Geography*, vol.20, pp.17-42.

Okafor, J. C. (1989), *Agroforestry Aspects*, WWF-UK/ODNRI, Godalming.

Okali, D.U.U. (1989), *Forestry Studies in conjunction with the Soil Survey and Land Evaluation of the Oban Division of the proposed Cross River National Park*, WWF-UK/ODNRI, Godalming.

Okali, D.U.U. (1990), *Forestry Aspects*, WWF-UK, Godalming.

Okali, D.U.U (1992), 'Funding needs and sources for a rain forest conservation project: the Cross River National Park Project', in J.A.T. Ojo (ed), *Proceedings on Mobilising Finances for Natural Resources Conservation in Nigeria*, NARESCON, Abuja, pp.180-183.

Okoye, J.C. (1988), 'Social clubs, resource management and rural development in Nigeria: lessons from Anambra State', *Transactions Institute of British Geographers* NS, vol.13, pp. 222-227.

Okurume, G.E. (1993), 'Agriculture', in I.A Adalemo and J.M. Baba (eds), *Nigeria: Giant in the Tropics, Vol. 1 - A Compendium*, Gabumo Publishing, Lagos, pp. 117-121.

Omoluabi, A.C. (1994), *Trade in Timber and Non-timber Forest Products*, Cross River State Forestry Project (ODA Assisted) Working Paper no. 6. Forestry Department Headquarters, Calabar.

Omoluabi, A.C. and Abang, S.O. (1994), *Marketing Margins in Non-timber Forest Products Trade in Cross River State of Nigeria*, Cross River State Forestry Project (ODA Assisted) Working Paper no. 12. Forestry Department Headquarters, Calabar.

Osemeobo, G.J. (1988), 'The human causes of forest depletion in Nigeria', *Environmental Conservation*, vol.15, pp. 18-28.

Osemeobo, G.J. (1990), 'Land use policies and biotic conservation: problems and prospects for forestry development in Nigeria', *Land Use Policy*, vol.7, pp. 314-322.

Osemeobo, G.J. (1991), 'Tenurial issues on natural resources conservation in Nigerian rainforest ecosystems', in F.A. Akinsanmi (ed), *Role of Forestry in Stabilising Fragile Ecosystems of the Rainforest Zone of Nigeria*, Proceedings of the 21st Annual Conference of the Forestry Association of Nigeria, Uyo, Akwa Ibom State, 7-12 April, 1992.

Osemeobo, G.J. (1993a), 'Agricultural land use in Nigerian forestry reserves: towards a solution to problems or conflict in biotic conservation', *Land Use Policy*, vol.10, pp. 44-48.

Osemeobo, G.J. (1993b), 'The hazards of rural poverty: decline in common property resources in Nigerian rainforest ecosystems', *Journal of Environmental Management*, vol.38, pp. 201-212.

Park, C.C. (1992), *Tropical Rainforests*, Routledge, London.

Pearce, D.W. and Turner, R.K. (1990), *Economics of Natural Resources and the Environment*, Harvester Wheatsheaf, New York and London.

Petters, S.W. (1993), 'Cross River State', R.K. Udo and A.B. Mamman (eds), *Nigeria: Giant in the Tropics*, Vol. 2, Heritage Edition, Gabumo Publishing, Lagos, pp.126-136.

Poore, D. and Sayer, J. (1987), *The Management of Tropical Moist Forest Lands: Ecological Principles*, IUCN, Gland, Switzerland.

Poore, D., Burgess, P., Palmer, J., Rietbergen, S. and Synnot, T. (1989), *No Timber Without Trees: Sustainability in the Tropical Forest*, Earthscan Publishers Ltd, London.

Repetto, R. (1988), *The Forest for the Trees ?: Government Policies and the Misuse of Forest Resources*, World Resources Institute.

Richards, P. (1985), *Indigenous Agricultural Revolution: Ecology and Food Production in West Africa*, Unwin Hyman, London.

Rock, M.T. (1996), 'The stork, the plow, rural societies and tropical deforestation in poor countries?', *Ecological Economics*, vol.18, pp.113-131.

Rubinoff, I. (1983), 'A strategy for preserving tropical forests', S.L. Sutton, T.C. Whitmore and A.C. Chadwick (eds), *Tropical Rain Forest: Ecology and Management*, Blackwell Scientific Publishers, pp. 465-476.

Rudel, T. (1994), 'Population, development and tropical deforestation: a cross-national study', in K. Brown and D. Pearce (eds), *The Causes of Tropical Deforestation*, University of London Press, 96-105.

Rudel, T.K. and Horowitz, B. (1993), *Tropical Deforestation: Small Farmers and Land Clearing in the Ecuadorian Amazon*, Columbia University Press, New York.

Ruthenberg (1976), *Farming Systems in the Tropics*, 2nd Edition, Clarendon Press, Oxford.

Sayer, J.C. (1991a), 'Buffer zones in rainforests; fact or fantasy ?', *Parks*, vol.2, pp. 20-24.

Sayer, J.C. (1991b), *Rainforest Buffer Zones: Guidelines for Protected Area Management*, IUCN, Gland, Switzerland.

Schelhas, J. (1991), 'A methodology for assessment of external issues facing National Parks with an application in Costa Rica', *Environmental Conservation*, vol.18 pp. 323-330.

Schelhas, J. (1992), 'Socio-economic and biological analysis for buffer zone establishment', *Developments in Landscape Management and Urban Planning*, vol.7, pp. 163-169.

Schelhas, J. and Shaw, W.W. (1992), 'Understanding Land Use and Natural Resource Decisions of Park Neighbours', Paper presented at the IV World Congress on National Parks and Protected Areas, Caracas, Venezuela, February.

Schonewald-Cox, C.M. and Bayless, J.W. (1986), 'The boundary model: a geographical analysis of design and conservation of nature reserves', *Biological Conservation*, vol. 38, pp. 305-322.

Schreiber, G. (1993), 'Nigeria: Proposed Third Forestry Project Pre-Appraisal Mission, March 14 - April 8, 1993', Aide Memoire, World Bank, Washington D.C.

Shafer, C.L. (1990), *Nature Reserves: Island Theory and Conservation Practice*, Smithsonian Press, Washington.

Shafer, C.L. (1999), 'National park and reserve planning to protect biological diversity: some basic elements', *Landscape and Urban Planning*, vol.44, pp. 123-153.

Sharma, U.R. (1990), 'An overview of people-park interactions in Royal Chitwan National Park, Nepal', *Landscape and Urban Planning*, vol.19, pp. 133-144.

Shuerholz, G., O. Ojong and P. Bisong (1990), *Okwangwo Division Management Plan*, WWF-UK, Godalming.

Sierra, R. (2000), 'Dynamics and patterns of Deforestation in the Western Amazon: the Napo deforestation front, 1986-1996', *Applied Geography*, vol.20, pp. 1-16.

Simmons, I.G. (1981), *The Ecology of Natural Resources*, Edward Arnold, London.

Singh, A. (1986), 'Farming Systems in Cross River State, 1985-86'. Faculty of Agriculture, University of Calabar, Mimeo.

Sommer, A. (1976), 'An attempt at an assessment of the world's tropical moist forests', *Unasylva*, vol.28, pp. 5-24.

Songorwa, A.N. (1999), 'Community-based wildlife management (CWM) in Tanzania: are communities interested?', *World Development*, vol.27, pp. 2061-2079.

Stocking, M. and Perkin, S. (1992), 'Conservation-with-development: an application of the concept in the Usambara Mountains, Tanzania', *Transactions, Institute of British Geographers NS*, vol.17, pp. 337-349.

Stonich, S.C. (1989), 'The dynamics of social processes and environmental destruction: a central American case study', *Population and Development Review*, vol.15, pp. 269-296.

Stuart, S.N., Adams, R.J. and Jenkins, M.D. (1990), *Biodiversity in Sub-Saharan Africa and its Islands: Conservation, Management and Sustainable Use*, Occasional Papers, IUCN Species Survival Commission, No. 6. IUCN, Gland, Switzerland.

Tomich, T.P., van Noordwijk, M., Vosti, S.A. and Witcover, J. (1998), 'Agricultural development with rainforest conservation: methods for seeking best bet alternatives to slash-and-burn, with applications to Brazil and Indonesia', *Agricultural Economics*, vol.19, pp.159-174.

Tufuor, K. (1992), 'Cross country issues in forest conservation and management', in K. Cleaver, M.Munasinghe, M. Dyson, N. Egli, A. Peuker and F. Wencelius (eds), *Conservation of West and Central African Rainforests*, World Bank Environment Paper No. 1, Washington D.C., pp. 53-62.

Turton, D. (1987), 'The Masaai and national park development in the Lower Omo Valley', in D. Anderson and R. Grove (eds), *Conservation in Africa: People, Policies and Practice*, Cambridge University Press, pp. 169-186.

Udo, R.K. (1990), *Land Use Policy and Land Ownership in Nigeria*, Ebieakwa Ventures, Lagos.

Udo, R.K. (1993), 'The population of Nigeria', in I.A. Adalemo and J.M. Baba (eds), *Nigeria: Giant in the Tropics, Vol. 1, A Compendium*, Gabumo Publishing, Lagos, pp. 29-33.

Umeh, L.I. (1992), *Forest Management in Nigeria: Problems and the Needed Strategies*, FORMECU, Ibadan.

Umeh, L.I. and Harou, P.A. (1992), 'Forestry resources issues and elements of a strategy for Nigeria', in K. Cleaver, M.Munasinghe, M. Dyson, N. Egli, A. Peuker and F. Wencelius (eds), *Conservation of West and Central African Rainforests*, World Bank Environment Paper No. 1, Washington D.C., pp 44-52.

Usani, U.O. (1992a), Cross River National Park (Okwangwo Division) Mbe Mountains Support Zone Development Activities: Quarterly Report, June 1992, WWF-CRNPP, Calabar.

Usani, U.O. (1992b), Cross River National Park (Okwangwo Division) Mbe Mountains Support Zone Development Activities: Quarterly Report, December 1992, WWF-CRNPP, Calabar.

Veldkamp, E., Weitz, A.M., Staristsky, I.G. and Huising, E.J. (1992), 'Deforestation trends in the Atlantic zone of Costa Rica: a case study', *Land Degradation and Rehabilitation*, vol.3, pp. 71-84.

World Commission on Environment and Development (WCED) (1987), *Our Common Future*, Oxford University Press, Oxford.

World Conservation Monitoring Centre (WCMC) (1988), Nigeria: Conservation of Biological Diversity. Unpublished Report, WCMC, Cambridge.

World Conservation Monitoring Centre (WCMC) (1992), *Global Diversity: Status of the Earth's Living Resources*, Chapman and Hall, London.

Weber, T. (1988), *Hugging the Trees*, Penguin.

Wells, M.P. and Brandon, K.E. (1993), 'The principles and practice of buffer zones and local participation in biodiversity conservation', *Ambio*, vol.22, pp. 157-162.

Wells, M., Brandon, K. and Hannah, L. (1992), *People and Parks: Linking Protected*

Areas Management and Local Communities, World Bank/WWF/USAID.

West, P.C and Brechin, S.R. (eds) (1991), *Resident Peoples and National Parks: Social Dilemmas and Strategies in International Conservation*, The University of Arizona Press, Tucson.

Western, D. (1982), 'Amboseli National Park: enlisting landowners to conserve migratory wildlife', *Ambio*, vol.11, pp. 302-308.

White, D. (1990), *Okwangwo Division Species Lists*, WWF-UK, Godalming.

Whitmore, T.C. (1984), *Tropical Rain Forests of the Far East*, 2nd Edition, Oxford University Press, Oxford.

Wickramasinghe, A. (1997), 'Anthropogenic factors and forest management in Sri Lanka', *Applied Geography*, vol.17, pp.87-110.

Williams, S.K.T. (1978), *Rural Development in Nigeria*, University of Ife Press, Ile-Ife.

World Bank (1992), *World Development Report 1992: Development and the Environment*, Oxford University Press.

WWF International (1991), *Tropical Forest Conservation*, Position Paper No. 7, Gland, Switzerland.

WWF-UK (1992), Boshi Okwangwo and Korup Coastal Rainforest Conservation Projects: Revised Proposal to the European Commission for Funding of Okwangwo Division of Cross River National Park, WWF Mimeo, Godalming.

Zube, E.H. (1986), 'Local and extra-local perceptions of national parks and protected areas', *Landscape and Urban Planning*, vol.13, pp. 11-17.

Zube, E.H. and Busch, M. (1990), 'Park-people relationships: an international review', *Landscape and Urban Planning*, vol.19, pp. 117-131.

Index